Capacitated Lot Sizing Problems in Process Industries

Ravi Ramya • Chandrasekharan Rajendran • Hans Ziegler
Sanjay Mohapatra • K. Ganesh

Capacitated Lot Sizing Problems in Process Industries

Springer

Ravi Ramya
Department of Management Studies
Indian Institute of Technology Madras
Chennai, TN, India

Chandrasekharan Rajendran
Department of Management Studies
Indian Institute of Technology Madras
Chennai, TN, India

Hans Ziegler
Chair of Production and Logistics
Universitát Passau
Passau, Bayern, Germany

Sanjay Mohapatra
Xavier Institute of Management
Bhubaneswar, Odisha, India

K. Ganesh
SCM Center of Competence, McKinsey Knowledge Center
McKinsey & Company
Chennai, TN, India

ISBN 978-3-030-01221-2 ISBN 978-3-030-01222-9 (eBook)
https://doi.org/10.1007/978-3-030-01222-9

Library of Congress Control Number: 2018958714

This Springer imprint is published by the registered company Springer Nature Switzerland AG
The registered company address is: Gewerbestrasse 11, 6330 Cham, Switzerland

Dedicated to
our parents and family members

Preface

Lot sizing is one of the most important decisions taken during production planning in all manufacturing and process industries. Among the various basic lot sizing models, the Capacitated Lot Sizing Problem (CLSP) is the main focus of this book. A single-machine, single-level and multiple-item CLSP is considered. It is the problem of planning the production of several items over a number of periods, satisfying all demand requirements so that the production times and setup times do not exceed the capacity limitations, and the total cost of the production plan is minimized. The main cost components are set up cost, inventory holding cost, backorder cost and lost-sales cost. The CLSP comes under the class of big-bucket lot sizing problems in the CLSP literature. It is considered as a big-bucket problem because several products/setups may be produced in a period. A period normally represents a time slot of one shift/day or one day or one week or one month (depending upon the production planner). For each lot produced in a period, a cost of setup occurs, and the time corresponding to the setup along with the production time (product of the production quantity and the number of units of a product produced per unit time considered) consumes the capacity in a period (assumed in time units). The CLSP is solved over a finite time horizon due to which excess quantity produced in a period can be stored to satisfy the demand of some future period, and the demand which cannot be met in a period due to capacity constraints can be backordered. Following the basic CLSP, studies on lot sizing consider the phenomenon where the setup state of a product is carried from one period to another in order to avoid multiple setups for the same product in consecutive periods. This phenomenon is called as setup carryover in the literature, whereas it is called production carryover in this book.

In almost all process industries, there are situations where some products have long and uninterrupted setup times, and the setup of the product and its consecutive production can be carried over across consecutive periods. Also, certain process industries require the production of a product to occur immediately after its setup, and the product to be continuously produced without any interruption. The phenomenon where the setup of a product having long setup time is carried over across periods is called setup crossover in the literature as well as in this book.

In this book, a mathematical model for the Capacitated Lot Sizing Problem with Production Carryover and Setup Crossover across periods (CLSP-PCSC), with possible backorders and with such real-life considerations in process industries, is proposed in Chap. 3. The aspect of allowing the setup to be carried over more than one period is called setup crossover in this book. The model proposed is all encompassing that it can handle continuous manufacturing (as in the case of process industries), and also situations where the setup costs and holding costs are product dependent and time independent/time dependent, with appropriate adaptations. The proposed model is also compared with an existing MILP (Mixed Integer Linear

Programming) model. A heuristic is also proposed to solve the CLSP-PCSC. The performance of the mathematical model and the heuristic is presented.

In Chap. 4, another mathematical model and a comprehensive heuristic are proposed for the same problem. This model is also all encompassing in that it can handle continuous manufacturing (as in process industries), and also situations where the setup costs and holding costs are product dependent and time independent/time dependent, with appropriate adaptations. A comprehensive heuristic is proposed based on this mathematical model to solve the CLSP-PCSC. The performance of the proposed model and the heuristic is evaluated using problem instances of various sizes.

In some process and manufacturing industries, the setup time of a machine not only depends on the time to setup a product but it also depends upon the product previously setup on the machine. This aspect where the setup time of a product on the machine depends on the time to setup a given product after setting up a given preceding product is called the sequence-dependent setup. The corresponding time taken for setting up the product is called as the sequence-dependent setup time and the corresponding cost involved is called the sequence-dependent setup cost. Researchers have considered the presence of sequence-dependent setup times and setup costs while addressing CLSP in industries in the presence of sequence-dependent setups. This book presents in Chap. 5 mathematical models developed for the Capacitated Lot Sizing Problem with Production Carryover and Setup Crossover across periods, assuming Sequence-Dependent Setup Times and Setup Costs (CLSP-SD-PCSC). In addition, these models allow the presence of backorders and also address real-life situations present in process industries such as the production of a product starting immediately after its uninterrupted setup and the uninterrupted production carryover across periods, along with the presence of long setup times. The consideration of these real-life situations is unique to this book and is one of its most significant contributions in every chapter.

In summary, this book addresses a class of CLSPs addressing some real-life situations present in process industries. Several variants of mathematical models and heuristics are proposed and developed to address some classes of lot sizing problems.

The authors, in particular the first author, gratefully acknowledges the support from Indian Institute of Technology Madras, University of Passau and German Academic Exchange Service (DAAD) for carrying out a major part of this work. The first author also thankfully acknowledges the comments from Prof. Rainer Leisten (University of Duisburg Essen, Germany) and Prof. Peeyush Mehta (Indian Institute of Management Calcutta) who had been her Ph.D. thesis examiners.

Chennai, India — Ravi Ramya
Chennai, India — Chandrasekharan Rajendran
Passau, Germany — Hans Ziegler
Bhubaneswar, India — Sanjay Mohapatra
Chennai, India — K Ganesh

Contents

LIST OF TABLES

LIST OF FIGURES

ABBREVIATIONS

APS	Advanced Planning Systems
B&B	Branch & Bound
BoM	Bill of Materials
CCO	Compressing Carry Over model
CLSD	Capacitated Lot Sizing Problem with Sequence-Dependent Setups
CLSP	Capacitated Lot Sizing Problem
CLSPL	Capacitated Lot Sizing Problem with Linked lot sizes
CLSP-PCSC	Capacitated Lot Sizing Problem with Production Carryover and Setup Crossover
CLSP-SD-PCSC	Capacitated Lot Sizing Problem with Sequence-Dependent Setups along with Production Carryover and Setup Crossover
CNC	Computer Numeric Control
CO	Carry Over model
CSLP	Continuous Setup Lot sizing Problem
DLSP	Discrete Lot sizing and Scheduling Problem
ELSP	Economic Lot Scheduling Problem
EOQ	Economic Order Quantity
EDI	Electronic Data Interchange
ERP	Enterprise Resource Planning
FL	Facility Location
FMS	Flexible Manufacturing System
GA	Genetic Algorithm
GRASP	Greedy Randomized Adaptive Search Procedure
LP	Linear Programming
LUC	Least Unit Cost
MES	Manufacturing Execution Systems
MILP	Mixed Integer Linear Programming
MIP	Mixed Integer Programming
MLCLSP	Multi-Level Capacitated Lot Sizing Problem
MM1:CLSP-PCSC	Mathematical Model 1 for the Capacitated Lot Sizing Problem with Production Carryover and Setup Crossover
MM2:CLSP-PCSC	Mathematical Model 2 for the Capacitated Lot Sizing Problem with Production Carryover and Setup Crossover
MM1:CLSP-SD-PCSC	Mathematical Model 1 for the Capacitated Lot Sizing Problem with Sequence-Dependent Setups along with Production Carryover and Setup Crossover

MM2:CLSP-SD-PCSC	Mathematical Model 2 for the Capacitated Lot Sizing Problem with Sequence-Dependent Setups along with Production Carryover and Setup Crossover
MRP	Material Requirements Planning
NCO	Non-Carry Over model
NP	Non-Polynomial
PLSP	Proportional Lot sizing and Scheduling Problem
POQ	Periodic Order Quantity
PPB	Part Period Balancing
SA	Simulated Annealing algorithm
SMC	Simple Multi-Commodity
SR	Shortest Route
SSR	Strengthened Shortest Route

Notations

N	Number of products
T	Number of time periods
t	Time period
i	Product
SC_i	Setup cost for product i
$SC_{i,t}$	Setup cost for product i, when its setup is initiated in period t; this cost is incurred only once as a fixed cost computed with respect to the period of its setup initiation
$SC^1_{i,t}$	Rate of cost of setup (cost corresponding to one time unit of setup) of product i in period t. It is given by $SC^1_{i,t} = \dfrac{SC'_{i,t}}{ST_i}$
$SC_{\phi,i}$	Sequence-dependent setup cost incurred in the machine for the first product i setup in period 1
$SC_{i',i}$	Sequence-dependent setup cost incurred when the machine is set up from product i' to product i
h_i	Holding cost per period per unit of product i
$h_{i,t}$	Holding cost per period per unit of product i in period t
b_i	Backorder cost per period per unit of product i
ST_i	Setup time for product i
$ST_{\phi,i}$	Sequence-dependent setup time for the first product i setup in the machine in period 1
$ST_{i',i}$	Sequence-dependent setup time when the machine is set up from product i' to product i
$ST'_{i,t}$	A variable that is assigned with the value of setup time of product i which is set up in period t
a_i	Number of time units required for producing one unit of product i
C_t	Capacity of the machine in period t (in time units)
$d_{i,t}$	Demand for product i in period t
M	A large value
$\mathcal{E}$	Smallest unit of time (a small positive real number)
$\mathcal{E}_d$	Unit of smallest quantity of production (a small positive real number)

$I_{i,t}$	Inventory of product i at the end of period t
$B_{i,t}$	Backorder quantity of product i at the end of period t

Definition of Variables Specific to Mathematical Model (MM1:CLSP-PCSC) in Chap. 3

Variable	Description
$\delta^1_{i,t}$	An indicator (binary) variable that takes value 1 if a complete setup is done for product i in period t with the production starting in period t; 0 otherwise
$\Delta^1_{i,t,t'}$	An indicator (binary) variable that takes value 1: it corresponds to the production carryover from period t to period t' ($t \leq t' \leq T$), due to the setup of product i started and finished in period t, with no intermittent setup of any other product; 0 otherwise
$\delta^2_{i,t}$	An indicator (binary) variable that takes value 1 if a setup of product i is started and completed exactly at the end of period t, followed by its production starting in period $t+1$; 0 otherwise
$\Delta^2_{i,t,t'}$	An indicator (binary) variable that takes value 1: it corresponds to the production carryover from period t' to period $t'+1$ ($t+1 \leq t' \leq T$), due to the end-of-period setup of product i in period t, with no intermittent setup of any other product; 0 otherwise
$\delta^3_{i,t,t'}$	An indicator (binary) variable that takes value 1 if the setup of product i is commenced in period t and is completed during some period t' but not exactly at the end of period t' ($t+1 \leq t' \leq T$); 0 otherwise

Variable	Description
$\Delta^3_{i,t,t',t''}$	An indicator (binary) variable that takes value 1: it corresponds to the production in period t'' ($t' \leq t'' \leq T$), due to the setup of product i initiated in period t and completed in period t' but not exactly at the end of period t' ($t+1 \leq t' \leq T$) and with no setup of any product during the intermittent periods from period t' to period t''. *Note:* This variable corresponds to (i.e. indicates) the production carryover through periods t' and t'', after the completion of setup in period t' (but not exactly at the end of period t'), with the initiation of setup in period t; 0 otherwise
$\delta^4_{i,t,t'}$	An indicator (binary) variable that takes value 1 if the setup of product i is commenced in period t and is completed exactly at the end of period t' ($t+1 \leq t' \leq T-1$); 0 otherwise
$\Delta^4_{i,t,t',t''}$	An indicator (binary) variable that takes value 1: it corresponds to the production in period t'' ($t'+1 \leq t'' \leq T$), due to the setup of product i initiated in period t and completed exactly at the end of period t' ($t+1 \leq t' \leq T-1$) and with no setup of any product during the intermittent periods from period t' to period t''. *Note:* This variable corresponds to (i.e. indicates) the production carryover through periods t' and t'', after the completion of setup exactly at the end of period t', with the initiation of setup in period t; 0 otherwise
$s^1_{i,t}$	Setup time of product i in period t that takes the value of ST_i, and associated with $\delta^1_{i,t}$
$s^2_{i,t}$	Setup time of product i in period t that takes the value of ST_i, and associated with $\delta^2_{i,t}$
$s^3_{i,t,t',t''}$	Setup time of product i in period t'' ($t \leq t'' \leq t'$), when its setup has started in period t and completed in period t', but not completed exactly at the end of period t', and associated with $\delta^3_{i,t,t'}$; *Note:* $\sum_{t''=t}^{t'} s^3_{i,t,t',t''} = ST_i$
$s^4_{i,t,t',t''}$	Setup time of product i in period t'' ($t \leq t'' \leq t'$), when its setup has started in period t and completed exactly at the end of period t', and associated with $\delta^4_{i,t,t'}$; *Note:* $\sum_{t''=t}^{t'} s^4_{i,t,t',t''} = ST_i$

Variable	Description
$X^1_{i,t,t'}$	Production quantity of product i in period t' (due to its setup started and ended in period t), with $1 \leq t \leq T$ and $t \leq t' \leq T$, and associated with $\Delta^1_{i,t,t'}$
$X^2_{i,t,t'}$	Production quantity of product i in period t' (due to its setup started in period t and completed exactly at the end of period t), with $1 \leq t \leq T-1$ and $t+1 \leq t' \leq T$, and associated with $\Delta^2_{i,t,t'}$
$X^3_{i,t,t',t''}$	Production quantity of product i in period t'' (due to its setup started in period t and ended in period t' but not completed at the end of period t'), with $1 \leq t \leq T-1$, $t+1 \leq t' \leq T$ and $t+1 \leq t'' \leq T$, and $t'' \geq t'$, and associated with $\Delta^3_{i,t,t',t''}$
$X^4_{i,t,t',t''}$	Production quantity of product i in period t'' (due to its setup started in period t and completed at the end of period t'), with $1 \leq t \leq T-2$, $t+1 \leq t' \leq T-1$ and $t+2 \leq t'' \leq T$, and $t'' > t'$, and associated with $\Delta^4_{i,t,t',t''}$

Definition of Variables Specific to Mathematical Model (MM2:CLSP-PCSC) in Chap. 4

Variable	Description
$\delta^1_{i,t}$	An indicator (binary) variable that takes value 1 if a complete setup is done for product i in period t with the production starting in period t; 0 otherwise
$\Delta^1_{i,t,t'}$	An indicator (binary) variable that takes value 1: it corresponds to a possible production carryover from period t to period t' ($t \leq t' \leq T$), due to the setup of product i started and finished in period t, with no intermittent setup of any other product; 0 otherwise
$\delta^2_{i,t}$	An indicator (binary) variable that takes value 1 if the setup of product i is started and completed exactly at the end of period t followed by its production starting in period $t+1$; 0 otherwise
$\Delta^2_{i,t,t'}$	An indicator (binary) variable that takes value 1: it corresponds to a possible production carryover from period t' to period $t'+1$ ($t+1 \leq t' \leq T$), due to the end-of-period setup of product i in period t, with no intermittent setup of any other product; 0 otherwise

Variable	Description
$\delta^3_{i,t}$	An indicator (binary) variable that takes value 1 if the setup of product i is commenced in period t and is carried over across periods, and is completed in some period t' ($t' = t+1, t+2, .., T$); 0 otherwise
$\Omega^3_{i,t,t'}$	An indicator (binary) variable that takes value 1 if the setup of product i is initiated in period t and is present in period t' ($t \leq t' \leq T$), with the setup of product i ending in a period later than period t, but not exactly at the end of that period; 0 otherwise
$\Omega^4_{i,t,t'}$	An indicator (binary) variable that takes value 1 if the setup of product i is initiated in period t and is present in period t' ($t \leq t' \leq T-1$), with the setup of product i ending in a period later than period t and setup getting completed exactly at the end of that period; 0 otherwise
$\Delta^3_{i,t,t'}$	An indicator (binary) variable that takes value 1: it corresponds to a possible production carryover from period t' to period $t'+1$ ($t+1 \leq t' \leq T$), with the setup of product i (having started in period t) ending in a period later than period t, but not exactly at the end of that period, and with no intermittent setup of any product during the production of product i; 0 otherwise
$\Delta^4_{i,t,t'}$	An indicator (binary) variable that takes value 1: it corresponds to a possible production carryover from period t' to period $t'+1$ ($t+2 \leq t' \leq T$), with the setup of product i (having started in period t) ending in a period later than period t and setup getting completed exactly at the end of that period, and with no intermittent setup of any product during the production of product i; 0 otherwise
$s^1_{i,t}$	Setup time of product i in period t that takes the value of ST_i, and associated with $\delta^1_{i,t}$
$s^2_{i,t}$	Setup time of product i in period t that takes the value of ST_i, and associated with $\delta^2_{i,t}$
$s^3_{i,t,t'}$	Setup time of product i in period t' due to its setup started in period t, and associated with $\Omega^3_{i,t,t'}$

Variable	Description
$s^4_{i,t,t'}$	Setup time of product i in period t' due to its setup started in period t, and associated with $\Omega^4_{i,t,t'}$
$X^1_{i,t,t'}$	Production quantity of product i in period t' (due to its setup starting and ending within period t), with $1 \leq t \leq T$ and $t \leq t' \leq T$, and associated with $\Delta^1_{i,t,t'}$
$X^2_{i,t,t'}$	Production quantity of product i in period t' (due to its setup started in period t and completed exactly at the end of that period), with $1 \leq t \leq T-1$ and $t+1 \leq t' \leq T$, and associated with $\Delta^2_{i,t,t'}$
$X^3_{i,t,t'}$	Production quantity of product i in period t' (due to its setup starting in period t and ending in a later period, but not at the end of that period), with $1 \leq t \leq T-1$ and $t+1 \leq t' \leq T$, and associated with $\Delta^3_{i,t,t'}$

Definition of Variables Specific to Mathematical Model 1 (MM1:CLSP-SD-PCSC) in Chap. 5

Variable	Description
$\delta^1_{i,t}$	An indicator (binary) variable that takes value 1 if a complete setup is done for product i in period t with the production starting in period t; 0 otherwise
$\Delta^1_{i,t,t'}$	An indicator (binary) variable that takes value 1: it corresponds to a possible production carryover from period t to period t' ($t \leq t' \leq T$), due to the setup of product i started and finished in period t, with no intermittent setup of any other product; 0 otherwise
$\delta^2_{i,t}$	An indicator (binary) variable that takes value 1 if the setup of product i is started and completed exactly at the end of period t, followed by its production starting in period $t+1$; 0 otherwise
$\Delta^2_{i,t,t'}$	An indicator (binary) variable that takes value 1: it corresponds to a possible production carryover from period t' to period $t'+1$ ($t+1 \leq t' \leq T$), due to the end-of-period setup of product i in period t, with no intermittent setup of any other product; 0 otherwise

Variable	Description
$\delta^3_{i,t}$	An indicator (binary) variable that takes value 1 if the setup of product i is commenced in period t and is carried over across periods, and is completed in some period t' ($t' = t+1, t+2, .., T$); 0 otherwise
$\Omega^3_{i,t,t'}$	An indicator (binary) variable that takes value 1 if the setup of product i is initiated in period t and is present in period t' ($t \le t' \le T$), with the setup of product i ending in a period later than period t, but not exactly at the end of that period; 0 otherwise
$\Omega^4_{i,t,t'}$	An indicator (binary) variable that takes value 1 if the setup of product i is initiated in period t and is present in period t' ($t \le t' \le T-1$), with the setup of product i ending in a period later than period t and setup getting completed exactly at the end of that period; 0 otherwise
$\Delta^3_{i,t,t'}$	An indicator (binary) variable that takes value 1: it corresponds to a possible production carryover from period t' to period $t'+1$ ($t+1 \le t' \le T$), with the setup of product i (having started in period t) ending in a period later than period t, but not exactly at the end of that period, and with no intermittent setup of any product during the production of product i; 0 otherwise
$\Delta^4_{i,t,t'}$	An indicator (binary) variable that takes value 1: it corresponds to a possible production carryover from period t' to period $t'+1$ ($t+2 \le t' \le T$), with the setup of product i (having started in period t) ending in a period later than period t and setup getting completed exactly at the end of that period, and with no intermittent setup of any product during the production of product i; 0 otherwise
$s^1_{i,t}$	Setup time of product i in period t that takes the value of $ST'_{i,t}$, and associated with $\delta^1_{i,t}$
$s^2_{i,t}$	Setup time of product i in period t that takes the value of $ST'_{i,t}$, and associated with $\delta^2_{i,t}$
$s^3_{i,t,t'}$	Setup time of product i in period t' that takes the value of $ST'_{i,t}$ due to its setup started in period t, and associated with $\Omega^3_{i,t,t'}$

Variable	Description
$s^4_{i,t,t'}$	Setup time of product i in period t' that takes the value of $ST'_{i,t}$ due to its setup started in period t, and associated with $\Omega^4_{i,t,t'}$
$X^1_{i,t,t'}$	Production quantity of product i in period t' (due to its setup starting and ending within period t), with $1 \leq t \leq T$ and $t \leq t' \leq T$, and associated with $\Delta^1_{i,t,t'}$
$X^2_{i,t,t'}$	Production quantity of product i in period t' (due to its setup started in period t and completed exactly at the end of that period), with $1 \leq t \leq T-1$ and $t+1 \leq t' \leq T$, and associated with $\Delta^2_{i,t,t'}$
$X^3_{i,t,t'}$	Production quantity of product i in period t' (due to its setup starting in period t and ending in a later period but not at the end of that period), with $1 \leq t \leq T-1$ and $t+1 \leq t' \leq T$, and associated with $\Delta^3_{i,t,t'}$
$X^4_{i,t,t'}$	Production quantity of product i in period t' (due to its setup starting in period t and ending at the end of a later period), with $1 \leq t \leq T-2$ and $t+2 \leq t' \leq T$, and associated with $\Delta^4_{i,t,t'}$
$del_{\phi,i,1}$	An indicator (binary) variable which tracks product i that has been setup with respect to the beginning of time period 1
$del_{0,i,t}$	An indicator (binary) variable which tracks product i that has been setup last, with respect to the beginning of time period t
$del_{i,0,t}$	An indicator (binary) variable which tracks product i that has been setup last, with respect to the end of time period t
$\xi'_{\phi,i,1}$	An indicator (binary) variable which indicates the initiation of setup of product i as the first setup in period 1
$\xi_{i',i,t}$	An indicator (binary) variable which indicates the setup of product i in period t, with the setup of both products i' and i being initiated in this period. Therefore it is evident that at least one of $\xi_{i',i,t}$ cannot exist in period t without being preceded by $\xi'_{i'',i',t}$. In other words, the first possible setup initiated in this period is denoted by $\xi'_{i'',i',t}$ (in general followed by subsequent possible initiated setups denoted by $\xi_{i',i,t}$)
$\xi'_{i,i',t}$	An indicator (binary) variable which indicates the initiation of setup of product i' as the first setup in period t, with the setup of product i' preceded by the setup of product i

Variable	Description
$delta_t$	An indicator (binary) variable which takes the value 1, if the setup of at least one of any product is initiated in period t (indicated by $delta_t = 1$). It also means that no carryover of either setup or production of any product, with its setup having been initiated in any period up to period $t-1$, can be carried over to period $t+1$ and later. This is so because if any product's setup is initiated in period t, then either the production or setup carryover with the setup initiated up to period $t-1$ which can be carried over to period $t+1$ and later is not feasible
$U_{i,t}$	An auxiliary variable that assigns product i in period t; associated with $\xi_{i',i,t}$ to avoid sub-tours
$ST'_{i,t}$	A variable that is assigned the value of setup time of product i which is initiated for its setup in period t

Definition of Variables Specific to Mathematical Model 2 (MM2:CLSP-SD-PCSC) in Chap. 5

Variable	Description
$\delta^1_{0,i,t}$	An indicator (binary) variable that takes value 1 if a complete setup is done for product i in the beginning of period $t = 1$ (as the first setup) with the production starting in period 1; 0 otherwise
$\delta^1_{i',i,t}$	An indicator (binary) variable that takes value 1 if a complete setup is done for product i (after setting up product i') in the beginning of period t with the production starting in period t; 0 otherwise
$\Delta^1_{i,t,t'}$	An indicator (binary) variable that takes value 1: it corresponds to the production carryover from period t to period t' ($t \leq t' \leq T$), due to the setup of product i started and finished in the beginning of period t, with no intermittent setup of any other product; 0 otherwise

Variable	Description
$\delta^2_{i',i,t}$	An indicator (binary) variable that takes value 1 if a setup is done for product i (after setting up product i') anywhere in the middle of period t except in the beginning or the end of period t followed by its production starting in period t; 0 otherwise
$\Delta^2_{i,t,t'}$	An indicator (binary) variable that takes value 1: it corresponds to the production carryover from period t to period t' ($t \leq t' \leq T$), due to the setup of product i started and finished anywhere in the middle of period t, with no intermittent setup of any other product; 0 otherwise
$\delta^3_{i',i,t}$	An indicator (binary) variable that takes value 1 if a setup is done for product i (after setting up product i') at the end-of-period t followed by its production starting in period $t+1$; 0 otherwise
$\Delta^3_{i,t,t'}$	An indicator (binary) variable that takes value 1: it corresponds to the production carryover from period t' to period $t'+1$ ($t+1 \leq t' \leq T$), due to the setup of product i started and completed exactly at the end-of-period t, with no intermittent setup of any other product; 0 otherwise
$\delta^4_{0,i,t}$	An indicator (binary) variable that takes value 1 if a complete setup is done for product i in period $t = 1$ (as the first setup) with the setup started exactly in the beginning of period t, and completed exactly at the end of period t, and with the production starting in period $t+1$. The setup time of the product setup using this indicator binary variable should be equal to the capacity of period 1; 0 otherwise
$\delta^4_{i',i,t}$	An indicator (binary) variable that takes value 1 if a complete setup is done for product i in period $t = 2, 3, .., T-1$ with the setup started exactly in the beginning of period t, and completed exactly at the end of period t, and with the production starting in period $t+1$. The setup time of the product setup using this indicator binary variable should be equal to the capacity of period t; 0 otherwise

Variable	Description
$\Delta^4_{i,t,t'}$	An indicator (binary) variable that takes value 1: it corresponds to the production carryover from period t' to period $t'+1$ ($t+1 \le t' \le T$), due to the complete setup of product i in period t (with the setup started exactly in the beginning of period t, and completed exactly at the end of period t), and with no intermittent setup of any other product; 0 otherwise
$\delta^5_{i',i,t,t'}$	An indicator (binary) variable that takes value 1 if the setup of product i is commenced in the beginning of period t and is completed during the middle of some period t' but not at the end of period t' ($t+1 \le t' \le T$), followed by its production starting in period t'; 0 otherwise
$\Delta^5_{i,t,t',t''}$	An indicator (binary) variable that takes value 1: it corresponds to the production in period t'' ($t' \le t'' \le T$), due to the setup of product i initiated in the beginning of period t and completed in the middle of period t' but not at the end of period t' ($t+1 \le t' \le T$), and with no setup of any product during the intermittent periods from period t' to period $t'' > t'$. *Note:* This variable corresponds to (i.e. indicates) the production carryover through periods t' and $t'' > t'$, after the completion of setup in the middle of period t', with the initiation of setup in the beginning of an earlier period t; 0 otherwise
$\Omega^5_{i,t,t',t''}$	An indicator (binary) variable that takes value 1 if the setup of product i is initiated in the beginning of period t and gets completed in the middle of a later period t' ($t+1 \le t' \le T$) but not exactly at the end of that period, with the setup of that product being present in period t'' ($t \le t'' \le t'$); 0 otherwise
$\delta^6_{i',i,t,t'}$	An indicator (binary) variable that takes value 1 if the setup of product i is commenced in the beginning of period t and is completed exactly at the end of period t' ($t+1 \le t' \le T-1$), followed by its production starting in period $t'+1$; 0 otherwise

Variable	Description
$\Delta^6_{i,t,t',t''}$	An indicator (binary) variable that takes value 1: it corresponds to the production in period t'' ($t'+1 \leq t'' \leq T$), due to the setup of product i initiated in the beginning of period t and completed exactly at the end of period t' ($t+1 \leq t' \leq T-1$), and with no setup of any product during the intermittent periods from period $t'+1$ to period $t'' > t'+1$. *Note:* This variable corresponds to (i.e. indicates) the production carryover through periods $t'+1$ and $t'' > t'+1$, after the completion of setup exactly at the end of period t', with the initiation of setup in the beginning of an earlier period t; 0 otherwise
$\Omega^6_{i,t,t',t''}$	An indicator (binary) variable that takes value 1 if the setup of product i is initiated in the beginning of period t and gets completed exactly at the end of period t' ($t+1 \leq t' \leq T-1$), with the setup of that product being present in period t'' ($t \leq t'' \leq t'$); 0 otherwise
$\delta^7_{i',i,t,t'}$	An indicator (binary) variable that takes value 1 if the setup of product i is commenced in the middle of period t and is completed during the middle of some period t' but not exactly at the end of period t' ($t+1 \leq t' \leq T$), followed by its production starting in period t'; 0 otherwise
$\Delta^7_{i,t,t',t''}$	An indicator (binary) variable that takes value 1: it corresponds to the production in period t'' ($t' \leq t'' \leq T$), due to the setup of product i initiated in the middle of period t and completed in the middle of period t' but not exactly at the end of period t' ($t+1 \leq t' \leq T$), and with no setup of any product during the intermittent periods from period t' to period $t'' > t'$. *Note:* This variable corresponds to (i.e. indicates) the production carryover through periods t' and $t'' > t'$, after the completion of setup in the middle of period t', with the initiation of setup in the middle of an earlier period t; 0 otherwise

Variable	Description
$\Omega^7_{i,t,t',t''}$	An indicator (binary) variable that takes value 1 if the setup of product i is initiated in the middle of period t and gets completed in the middle of a later period t' ($t+1 \leq t' \leq T$) but not exactly at the end of that period, with the setup of that product being present in period t'' ($t \leq t'' \leq t'$); 0 otherwise
$\delta^8_{i',i,t,t'}$	An indicator (binary) variable that takes value 1 if the setup of product i is commenced in the middle of period t and is completed exactly at the end of period t' ($t+1 \leq t' \leq T-1$), followed by its production starting in period $t'+1$; 0 otherwise
$\Delta^8_{i,t,t',t''}$	An indicator (binary) variable that takes value 1: it corresponds to the production in period t'' ($t'+1 \leq t'' \leq T$), due to the setup of product i initiated in the middle of period t and completed exactly at the end of period t' ($t+1 \leq t' \leq T-1$), and with no setup of any product during the intermittent periods from period $t'+1$ to period $t'' > t'+1$. *Note:* This variable corresponds to (i.e. indicates) the production carryover through periods $t'+1$ and $t'' > t'+1$, after the completion of setup at the end of period t', with the initiation of setup in the middle of an earlier period t; 0 otherwise
$\Omega^8_{i,t,t',t''}$	An indicator (binary) variable that takes value 1 if the setup of product i is initiated in the middle of period t and gets completed in the middle of a later period t' ($t+1 \leq t' \leq T-1$), with the setup of that product being present in period t'' ($t \leq t'' \leq t'$); 0 otherwise
$s^1_{i,t}$	Setup time of product i that takes the value of $ST_{0,i}$ when associated with $\delta^1_{0,i,t}$, in period $t=1$; and setup time of product i that takes the value of $ST_{i',i}$ when associated with $\delta^1_{i',i,t}$, in periods $t = 2, 3, .., T$
$s^2_{i,t}$	Setup time of product i in period t that takes the value of $ST_{i',i}$, and associated with $\delta^2_{i',i,t}$
$s^3_{i,t}$	Setup time of product i in period t that takes the value of $ST_{i',i}$, and associated with $\delta^3_{i',i,t}$

Variable	Description
$s^4_{i,t}$	Setup time of product i that takes the value of $ST_{0,i}$ when associated with $\delta^4_{0,i,t}$, in period $t = 1$; and setup time of product i that takes the value of $ST_{i',i}$ when associated with $\delta^4_{i',i,t}$, in periods $t = 2, 3, .., T - 1$
$s^5_{i,t,t',t''}$	Setup time of product i in period t'' ($t \leq t'' \leq t'$), when its setup has started in the beginning of period t and completed in the middle of period t' ($t + 1 \leq t' \leq T$), but not completed exactly at the end of period t', and associated with $\delta^5_{i',i,t,t'}$; *Note:* $\sum_{t''=t}^{t'} s^5_{i,t,t',t''} = ST_{i',i}$
$s^6_{i,t,t',t''}$	Setup time of product i in period t'' ($t \leq t'' \leq t'$), when its setup has started in the beginning of period t and completed exactly at the end of period t' ($t+1 \leq t' \leq T-1$), and associated with $\delta^6_{i',i,t,t'}$; *Note:* $\sum_{t''=t}^{t'} s^6_{i,t,t',t''} = ST_{i',i}$
$s^7_{i,t,t',t''}$	Setup time of product i in period t'' ($t \leq t'' \leq t'$), when its setup has started in the middle of period t and completed in the middle of period t' ($t + 1 \leq t' \leq T$), and associated with $\delta^7_{i',i,t,t'}$; *Note:* $\sum_{t''=t}^{t'} s^7_{i,t,t',t''} = ST_{i',i}$
$s^8_{i,t,t',t''}$	Setup time of product i in period t'' ($t \leq t'' \leq t'$), when its setup has started in the middle of period t and completed exactly at the end of period t' ($t + 1 \leq t' \leq T - 1$), and associated with $\delta^8_{i',i,t,t'}$; *Note:* $\sum_{t''=t}^{t'} s^8_{i,t,t',t''} = ST_{i',i}$
$X^1_{i,t,t'}$	Production quantity of product i in period t' (due to its setup started in the beginning of period t, and ended before the end of period t), with $1 \leq t \leq T$ and $t \leq t' \leq T$, and associated with $\Delta^1_{i,t,t'}$
$X^2_{i,t,t'}$	Production quantity of product i in period t' (due to its setup initiated in the middle of period t but not at the beginning or end of period t, and completed before the end of period t), with $1 \leq t \leq T$ and $t \leq t' \leq T$, and associated with $\Delta^2_{i,t,t'}$
$X^3_{i,t,t'}$	Production quantity of product i in period t' (due to its setup started and completed exactly at the end of period t), with $1 \leq t \leq T-1$ and $t+1 \leq t' \leq T$, and associated with $\Delta^3_{i,t,t'}$
$X^4_{i,t,t'}$	Production quantity of product i in period t' (due to its setup initiated at the beginning of period t and completed exactly at the end of period t, with the setup time of product i occupying the entire capacity of period t), with $1 \leq t \leq T - 1$ and $t + 1 \leq t' \leq T$, and associated with $\Delta^4_{i,t,t'}$

Variable	Description
$X^5_{i,t,t',t''}$	Production quantity of product i in period t'' (due to its setup started in the beginning of period t, and completed in the middle of period t' but not completed exactly at the end of period t'), with $1 \leq t \leq T-1$, $t+1 \leq t' \leq T$ and $t+1 \leq t'' \leq T$, and $t'' \geq t'$, and associated with $\Delta^5_{i,t,t',t''}$
$X^6_{i,t,t',t''}$	Production quantity of product i in period t'' (due to its setup started in the beginning of period t and completed exactly at the end of period t'), with $1 \leq t \leq T-2$, $t+1 \leq t' \leq T-1$ and $t+2 \leq t'' \leq T$, and $t'' > t'$, and associated with $\Delta^6_{i,t,t',t''}$
$X^7_{i,t,t',t''}$	Production quantity of product i in period t'' (due to its setup started in the middle of period t, and completed in the middle of period t' but not completed exactly at the end of period t'), with $1 \leq t \leq T-1$, $t+1 \leq t' \leq T$ and $t+1 \leq t'' \leq T$, and $t'' \geq t'$, and associated with $\Delta^7_{i,t,t',t''}$
$X^8_{i,t,t',t''}$	Production quantity of product i in period t'' (due to its setup started in the middle of period t and completed exactly at the end of period t'), with $1 \leq t \leq T-2$, $t+1 \leq t' \leq T-1$ and $t+2 \leq t'' \leq T$, and $t'' > t'$, and associated with $\Delta^8_{i,t,t',t''}$
$delta_t$	An indicator (binary) variable which takes the value 1, if the setup of at least one of any product is initiated in period t (indicated by $delta_t = 1$). It also means that no carryover of either setup or production of any product, with its setup having been initiated in any period up to period $t-1$, can be carried over to period $t+1$ and later. This is so because if any product's setup is initiated in period t, then either the production or setup carryover with the setup initiated up to period $t-1$ which can be carried over to period $t+1$ and later is not feasible
$U_{i,t}$	An auxiliary variable that assigns product i in period t; it helps to avoid subtours
$\alpha_{i,t}$	An indicator (binary) variable which tracks product i that has been setup last, with respect to the beginning of time period t
$\beta_{i,t}$	An indicator (binary) variable which tracks product i that has been setup last, with respect to the end of time period t

CHAPTER 1

Introduction

1.1 Preamble to the Production Planning Problem

In the initial phase of research in the fields of Operations Research and Management Science, researchers carried out their work on production planning under the assumption that demand was deterministic and time varying (dynamic). In due course, research on production planning evolved with the development of Harris's EOQ (Economic Order Quantity) formula, Wilsons's (Q,r) model, and the dynamic lot sizing model proposed by Wagner and Whitin. Harris (1913) was the first to publish the work in the lot sizing area entitled, (published as Harris (1990)). The EOQ was also known as Wilson's lot size formula as it was used in practice by Wilson (Wilson 1934). Wagner and Whitin developed the dynamic lot sizing problem in 1958 (Wagner and Whitin 1958). Since then, researchers deal with various types of lot sizing problems for different applications (also see Manne (1958)).

In order to take a major step forward in the standardization and control of production planning systems, Material Requirements Planning (MRP) was introduced in the 1970s. Although this major step was taken, criticism prevailed because it was very difficult to deal with lead time and capacity constraints within the existing system. Enterprise Resource Planning (ERP) and Advanced Planning Systems (APS) were not able to handle completely the complexity, when capacity constraints were involved.

Parallely, in the early 1960s and 1970s Mixed Integer Programming (MIP) problems were introduced to handle production planning. Since then various kinds of production planning problems have been solved using Mixed Integer Programming (MIP) models, which in turn are solved using standard solvers.

Production planning as defined by Pochet and Wolsey (2006) is the planning of acquisition of resources and raw materials, and the planning of production activities which require the transformation of raw materials into finished products, thereby meeting the customers demand in the most efficient and economic way possible. It can also be stated as the planning of future production given the resources available, such as man, machine and money; capacity on the machine available; and the amount raw material available when there is demand to be met for various products over a finite time horizon. In production planning, the best use of the production resources is carried out, thereby satisfying the production requirements, and anticipating sales opportunities (Karimi et al. 2003). The main goal of production planning is to make a trade-off between the economic objectives such as the cost minimization

R. Ramya et al., *Capacitated Lot Sizing Problems in Process Industries*, https://doi.org/10.1007/978-3-030-01222-9_1

or profit maximization, and satisfying the demands of the customers. To achieve this goal, the current trend to improve productivity is through supply chain coordination, wherein the production planning models are integrated with the models involving procurement, distribution and sales.

In production planning, three types of decisions can be made. They are: short term planning, medium term planning and long term planning (Karimi et al. 2003). Long term planning usually focuses on making strategic decisions involving the products which need to be produced, the processes which need to be chosen to produce the product, the choice of equipment's required to produce the product, the design and location of the facility which accommodates the resources, and planning of resources. Medium term planning involves decisions made regarding the quantity of product to be produced, the lot sizing decisions, and Material Requirements Planning (MRP), thereby optimizing the lot sizing problem by minimizing the cost or maximizing the profit, subject to the capacity constraints and satisfying the demand requirements. Short term planning involves decisions such as sequencing of products in a shop floor. The present work mainly focuses on medium term production planning decisions such as lot sizing.

Production planning is termed as "lot sizing" in a discrete production environment and "campaign planning" in a continuous production environment such as process industries (Suerie 2006). However, there seems to be no big difference between lot sizing and campaign planning. Other than the latter requiring some side constraints, the term "lot sizing" is mostly used in literature for production planning. There are various lot sizing problems such as ELSP (Economic Lot Scheduling Problem), DLSP (Discrete Lot sizing and Scheduling Problem), CSLP (Continuous Setup Lot sizing Problem), PLSP (Proportional Lot sizing and Scheduling Problem) and CLSP (Capacitated Lot Sizing Problem). This book focuses on the CLSP.

In this chapter, the attributes related to lot sizing are discussed in Sect. 1.2 and the classifications of various lot sizing problems are presented in Sect. 1.3. This is followed by the introduction to the lot sizing literature in Sect. 1.4. Here the literature on CLSP classified based on different categories is presented.

1.2 Basic Characteristics and Attributes of Lot Sizing Models

Lot sizing models are developed for planning the production of certain products, given the demand for those products over a finite time horizon. The characteristics or attributes of the lot sizing models should be known in order to determine the complexity of the lot sizing models (Karimi et al. 2003; Suerie 2006). There are two main characteristics/attributes which affect the modelling and complexity of lot sizing decisions. They are time based characteristics (Sect. 1.2.1) and product based characteristics (Sect. 1.2.2).

1.2.1 Time Based Characteristics and Attributes

The first set of characteristics and attributes are related to the time structure of the model developed, and the data used in making lot sizing decisions.

Planning Horizon

Lot sizing models are developed for planning the production in a planning horizon which may be finite or infinite. In a finite time horizon, the demand for the products may vary in every period (dynamic demand), nevertheless, in an infinite time horizon, normally a constant demand rate (stationary demand) is assumed similar to the EOQ model. While considering the planning horizon, a variant known as the rolling horizon is considered when there is uncertainty in data.

Time Scale

Continuous or Discrete time scale may be used in lot sizing problems. Lot sizing problems assuming a discrete time scale have big or small time buckets which may be uniform or non-uniform. Generally, uniform time buckets are assumed in standard lot sizing problems. The time buckets are classified into small and big time buckets, when the comparison of the relative length of the time period is made with respect to the relative length of the production lot (Salomon et al. 1991). Models assuming small time buckets allow at most one or two products to be produced in a period (due to this, sequence of the production orders can be easily determined) and models assuming big time buckets allow more than two products to be produced in a period (sequencing the production of products is difficult in big bucket models).

Parameters/Data

Some of the parameters required while modeling lot sizing problems are production capacity, cost parameters, production coefficients, setup time, etc. which may vary or not vary over time. Also, in deterministic models, the parameters are known prior to the planning of production. In stochastic models, an uncertainty of the future is incorporated in production planning.

Objective Function

The minimization of several cost components (such as setup cost, holding cost, backorder cost and lost sales cost) is the general objective of any lot sizing problem. Temporal objectives such as minimizing the maximum lateness or minimizing the total completion time are also used (Potts and van Wassenhove 1992) in lot sizing problems.

Cost Components

The cost components involved in lot sizing problems are setup cost, inventory holding cost, backorder cost and lost sales cost. Setup costs are incurred during the production process. Whenever a lot is produced, the resources have to be set up (e.g. retooling of the machine), to produce the next product. Direct costs involved in setup (e.g. cost of cleaning materials used for cleaning the machine) are also a part of the setup cost.

Capacity

The capacity of the resources is assumed to be either finite (capacitated) or infinite (uncapacitated). Resources having finite capacity can be extended overtime, associated with a certain cost in the lot sizing models. When capacity constraints exist, solving the lot sizing problem is very difficult. The capacity available in a period can be determined in time units. In some industries, the capacity of a resource in the current period is determined by the usage of the same resource in the former periods (Kimms 1997).

Number of Resources

The production planning problem can consider one or more resources. If a single operation needs two resources in parallel, they are termed to be “multiple resources”, and if an operation can be performed in more than one resource, they are termed as “parallel resources”.

Setup Operation

When a product is set up during lot sizing, a time known as “setup time” and a cost known as “setup cost” are associated with the setup operation. Setups can also be characterized as sequence-dependent setups in some industries. Minor and major setups are also distinguished in certain industries. Major setups are those which are done between products of different families, and minor setups are those which are done between products of the same family (Potts and van Wassenhove 1992).

1.2.2 Product Based Characteristics and Attributes

The second set of characteristics and attributes are related to the output of the production process.

Number of Products

There are lot sizing problems assuming production of a single product or multiple products. Also, when the Bill of Materials (BoM) structure is considered for a product, a product may have a single level or multiple levels. In products having a single level, the end items are directly produced from the raw materials with no intermediate sub-assemblies. Therefore, the demand for the single-level products can be known from forecasts or direct customer orders. In multi-level products, the raw materials after passing through several operations result in end products. Therefore, the output of one operation is the input to another operation, resulting in a condition where the demand of one level depends on the demand at another level. This type of demand is known as the dependent-demand. Solving multi-level lot sizing problems is more difficult than solving single-level lot sizing problems.

Inventory Restrictions

In lot sizing problems, there are upper bounds and lower bounds given for the inventory which is allowed to be held at the end of a certain period. The upper bound may be the capacity of the warehouse, and the lower bound may be the quantity of the safety stock to be held (Belvaux and Wolsey 2001). Shelf-life restrictions are imposed in inventory management, and are common in industries manufacturing perishable products such as food industries (Brown et al. 2001).

Service Policy

These policies concern the demand fulfilment. There are lot sizing models in which the unmet demand is lost (lost sales) or fulfilled in subsequent periods (by backordering) (Haase and Drexl 1994). Some models do not allow the presence of backorders or lost sales.

1.3 Classification of Lot Sizing Models

The lot sizing models are classified based on the "time scale" attribute of the lot sizing problems. Based on the time scale, the lot sizing problems can be classified as Continuous lot sizing problems and Dynamic lot sizing problems. A complete classification of lot sizing problems is explained in this section (see Fig. 1.1 for the classification of lot sizing problems).

1.3.1 Continuous Lot Sizing Problem: Economic Lot Scheduling Problem (ELSP)

In continuous lot sizing problem, the time scale considered is continuous and infinite (i.e. not divided into distinct periods). The classical economic lot scheduling problem (ELSP) comes under the class of Continuous Lot Sizing Problems.

The ELSP is a problem in operations management, which was first studied by Jack D. Rogers in 1958 in Berkeley. The problem is an extension of the EOQ model when several products are to be produced on the same machine. The classic ELSP is concerned with the production scheduling of several products on the same machine, in order to decide when to produce and how much to produce each product such that the setup cost and inventory holding costs are minimized. This problem is mainly studied to accommodate a cyclical production pattern of the individual items. This problem assumes stationary (constant) demand, an infinite planning horizon, and deals with the objective of developing a cyclic schedule which minimizes the average costs. The methods proposed for solving the ELSP was surveyed by Elmaghraby (1978). When capacity restrictions are involved, solving the ELSP is NP-hard.

1.3.2 Dynamic Lot Sizing Problem

The dynamic lot sizing problems deal with lot sizing decisions for single/multiple items when the demand is assumed to be deterministic and time varying. It is further classified into single-level or multi-level lot sizing problems. Single-level lot sizing problems are further classified into uncapacitated and capacitated lot sizing problems. The Wagner and Whitin algorithm and its extensions come under the uncapacitated lot sizing problem. The capacitated lot sizing problems are further classified into small bucket and big bucket lot sizing models. A brief explanation of the uncapacitated and capacitated lot sizing problems is given below.

Uncapacitated Lot Sizing Problem

The uncapacitated single-level lot sizing problem was addressed by Wagner and Whitin (1958), by assuming the demand to be dynamic and deterministic. They provided an iterative dynamic programming algorithm in order to compute the optimal ordering policy with no beginning inventory, and developed the "planning horizon theorem". Time-dependent cost structure was also incorporated into their model. The planning horizon results of Wagner and Whitin were improved to incorporate marginal production costs varying with time by Zabel (1964) and

Eppen et al. (1969). It was extended by Zangwill (1966) in order to allow the presence of backorders. Zangwill (1969) extended the model by incorporating multiple facilities. Since the algorithm proved to be inefficient to solve problems involving large number of products, several heuristics were proposed during 1960s and 1970s. Some of them were Lot-for-Lot, Periodic Order Quantity (POQ), Modified EOQ, Least Unit Cost (LUC) and Part Period Balancing (PPB). A heuristic based on average cost per period was proposed by Silver and Meal (1973). A worst case performance analysis for these heuristics was derived by Bitran et al. (1984). Bahl et al. (1987) addressed uncapacitated lot sizing problem in an uncertain environment. Wolsey (1989) developed a polynomial separation algorithm with start-up costs for the uncapacitated lot sizing problem. In order to reduce the computational complexity of the Wagner and Whitin algorithm, Wagelmans et al. (1992), Aggarwal and Park (1993) and Federgruen and Tzur (1991) proposed algorithms for the uncapacitated case of Wagner & Whitin dynamic lot sizing problem. Some solution approaches were also provided by Wolsey (1995).

Capacitated Lot Sizing Problem

The addition of capacity restrictions makes the lot sizing problems very hard to be solved using techniques like dynamic programming. The capacitated lot sizing problems are classified into small bucket and big bucket lot sizing models, based on the length of the time period.

Small Bucket Lot Sizing Models

In small bucket lot sizing models, the lot sizing and sequencing decisions are made simultaneously. Therefore, one or at most two setups can take place in a period. There are different types of small bucket models introduced in lot sizing literature. Some of the models are explained in this section.

(a) ***Discrete Lot sizing and Scheduling Problem:*** In a discrete lot sizing and scheduling problem (DLSP), several products can be produced in one shared resource which has a capacity limit. All the products face a deterministic and dynamic demand, and the planning horizon is finite (divided into T periods). Only one product can be produced in each period. If a product is produced in a period, it is produced at the full capacity during that period (all or nothing assumption for production). The setup of a product is associated with a setup cost, and the setup time is not reflected in DLSP modelling. The main objective of DLSP is to minimize the setup costs and holding costs of the products. The setup states can also be preserved across periods in DLSP, i.e., the production of a product i can continue from period $t-1$ to period t without performing a new setup for product i. This can happen when product i is set up and produced in period $t-1$, and the demand for product i exists in period t also. If the demand for product i is not present in period t, it can still be produced in that period to meet the demand in some later period by holding inventory (Fleischmann 1990; Magnanti and Vachani 1990).

(b) ***Continuous Setup Lot sizing Problem:*** The continuous setup lot sizing problem (CSLP) (Karmarkar and Schrage 1985; Salomon et al. 1991) has all its assumptions same as DLSP except the all or nothing assumption. Therefore, the lot size can take arbitrary values, i.e., the production of a product need not fill the entire period. Apart from this, the setup state can be preserved across idle periods, unlike DLSP.

(c) ***Proportional Lot sizing and Scheduling Problem:*** The proportional lot sizing and scheduling problem (PLSP) (e.g. Haase and Drexl (1994), Drexl and Haase (1995), Kimms and Drexl (1998), Belvaux and Wolsey (2001)) retains all the assumptions of the DLSP and CSLP except the all or nothing assumption of DLSP (i.e., only one product can be produced in a period). This assumption in CLSP is replaced by the assumption that two products can be produced in each period, i.e., one before and one after the setup operation. Since two products can be produced in a period in PLSP, the setup operation in a period is associated with a setup time so that the utilization of the capacity can be calculated in a period.

Big Bucket Lot Sizing Models

The capacitated lot sizing problem (CLSP) is a big bucket lot sizing model which is the focus of this present work. Initially the CLSP included only the presence of setup costs. The optimal production plan for multiple items with sequence-independent setup costs, and without any setup times, having capacity constraints for a single machine, is referred to as the capacitated lot sizing problem. Later, positive setup times were incorporated while developing mathematical models for CLSP.

When setup costs and setup times are assumed in CLSP, several items can be produced using a single resource which has a capacity limit. The planning horizon has a finite number of periods (T). The products face a dynamic and deterministic demand. If the products have to be set up in a certain period, the resource has to be set up for the product in that period. The setup of a product incurs a setup cost, and consumes a certain amount of capacity in the period in which the product is set up (given by the value of setup time). The main objective of the CLSP is to minimize the sum of the setup cost and holding cost of all products across all time periods. In CLSP, within each individual bucket, the sequence of the products cannot be determined. Only the partial sequence of the products can be known in their hybrid models which consider the setup state to be preserved across two periods. The assumptions and the basic mathematical formulation for the CLSP with setup costs and setup times are presented below (Suerie 2006).

Assumptions

- A single resource is considered in the problem in this book, and several products i are produced on the same resource.
- The resource has a finite capacity.
- Each product is made up of a single level and the products face a deterministic demand.
- Time unit is discrete and the time horizon considered is finite.
- When a product has to be produced in a certain period, the resource has to be set up for that product in the same period.
- The setups consume resource capacity.
- Each product is associated with a setup cost when set up on the machine, and it consumes time for setup.

- Backorders and lost sales are not permitted.
- The capacity of the machine during a given period is assumed in time units and it may vary from period to period.
- The capacity of the machine per period is consumed by the setup time and the production time of the products.
- The excess quantity produced of a product can be stored, and this incurs a holding cost, except in the last period where all the units in the inventory have to be consumed.
- The aim is to minimize the sum of holding costs and setup costs.

Notations
i: Products
t: Periods

Parameters
a_i: Production coefficient or the capacity consumption to produce one unit of product i
$b_{i,t}$: Large number, not limiting feasible production quantities of product i in period t
C_t: Available capacity in period t
$d_{i,t}$: Demand for product i in period t
$h_{i,t}$: Holding cost per unit of product i in period t
SC_i: Setup cost for product i
ST_i: Setup time for product i

Variables
$I_{i,t}$: Inventory of product i at the end of period t
$X_{i,t}$: Production quantity of product i in period t
$Y_{i,t}$: Binary decision variable which takes the value 1, if a setup for product i is performed in period t; 0 otherwise

Formulation
Objective Function
Minimize

$$\sum_{i=1}^{N}\sum_{t=1}^{T} SC_i Y_{i,t} + \sum_{j=1}^{N}\sum_{t=1}^{T} h_{i,t} I_{i,t} \tag{1.1}$$

subject to the following:

$$I_{i,t-1} + X_{i,t} = d_{i,t} + I_{i,t} \qquad \forall i \text{ and } \forall t. \tag{1.2}$$

$$\sum_{i=1}^{N} a_i X_{i,t} + \sum_{i=1}^{N} ST_i Y_{i,t} \leq C_t \qquad \forall t. \tag{1.3}$$

$$X_{i,t} \leq b_{i,t} Y_{i,t} \qquad \forall i \text{ and } \forall t. \tag{1.4}$$

$$X_{i,t} \geq 0,\ I_{i,t} \geq 0 \text{ and } I_{i,0} = 0 \qquad \forall i \text{ and } \forall t. \tag{1.5}$$

$$Y_{i,t} \in \{0, 1\} \qquad \forall i \text{ and } \forall t. \tag{1.6}$$

This mathematical formulation addresses the CLSP with setup costs and setup times. There are also various hybrid models of CLSP which have been vastly addressed in literature. Some of the hybrid models are presented in this section.

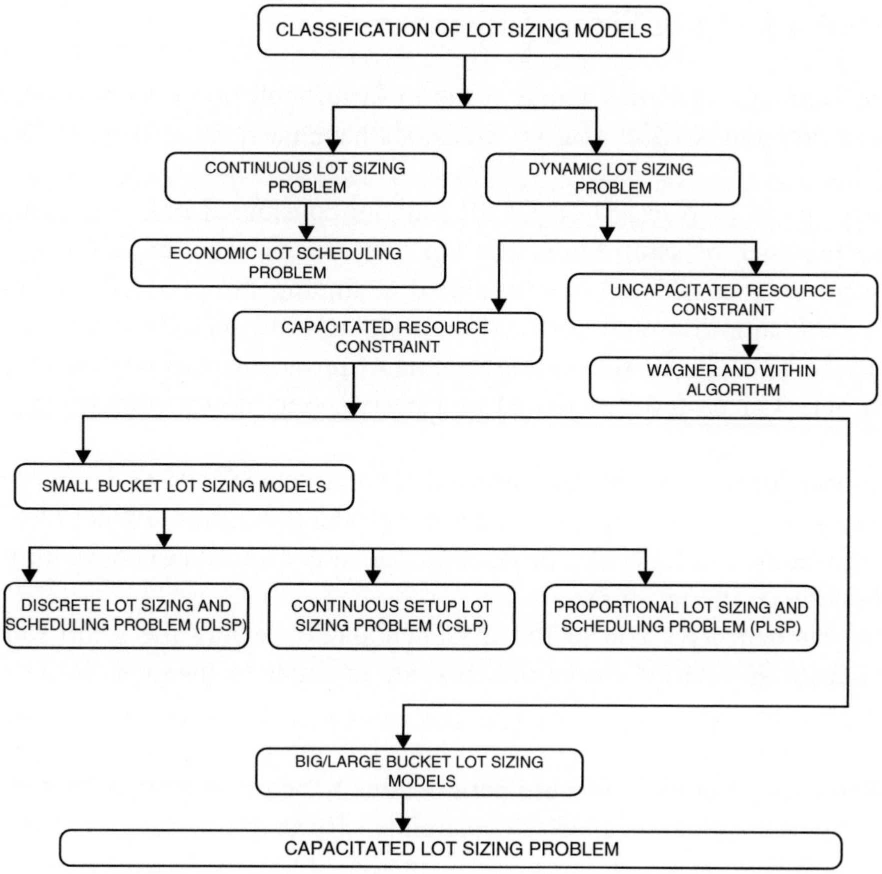

Figure 1.1: Classification of lot sizing models

(a) ***CLSP with production carryover across periods:*** Several researchers such as Dillenberger et al. (1994), Gopalakrishnan et al. (1995), Haase (1998), Suerie and Stadtler (2003), Caserta et al. (2009), Wu and Shi (2011), Caserta and Voß (2013), Mohapatra (2015a,b,c, 2017), and Sarkar et al. (2017) have addressed this problem where the setup state can be preserved across several periods, and have used various terms in their paper such as "linked lot sizes" and "setup carryover". In the present work which involves a class of capacitated lot sizing problems, the preservation of setup state across periods is called as "production carryover". Here, the production of product i can be carried over to period t from period $t-1$, by preserving the setup state of product i. The other assumptions of modelling are same as the basic CLSP with setup costs and setup times.

(b) ***CLSP with sequence-dependent setups:*** In this hybrid model, the CLSP assumes sequence-dependent setup costs and setup times (Mohapatra and Rajeev 2012; Mohapatra and Anuj 2010). Some of the authors who have assumed sequence-dependent setup times and setup costs are Haase (1996), Grünert (1998), Magnusson (2001), Eren and Güner (2006), Almada-Lobo et al. (2007), Mohapatra and Murarka (2016), Mohapatra et al. (2013), and Clark et al. (2014). The models developed by some of them assume production carryover across periods. When sequence-dependent setups are considered, the triangular inequality condition is assumed wherein the cost/time to set up product i'' after setting up to product i, is greater than the sum of cost/time incurred when product i' is set up after product i and product i'' is set up after product i'.

1.4 An Analysis of Lot Sizing Literature

In production planning, lot sizing is one of the most difficult problems to solve. The capacitated lot sizing problem (CLSP) and its extensions have been extensively studied by various researchers.

In the early stages of research on CLSP, authors considered only the presence of setup costs, and the presence of setup times was not considered. This basic CLSP in which the setup costs were only considered mainly aimed at finding the production plan on a single machine by minimizing the setup cost and the holding cost, thereby satisfying the capacity constraints. In the basic CLSP, scheduling (timing of production) of products within a period is not carried out. For each lot produced in a period, only a cost proportional to the setup occurs.

Later, authors addressing CLSP incorporated positive setup times, or a proportional amount of time corresponding to the setup along with the production time (product of the production quantity and the number of units of a product produced per unit time) consuming the capacity of the period (assumed in time units).

Further, researchers have considered the phenomenon where the setup state of a product is carried from one period to another period, in order to avoid multiple setups of the same product in consecutive periods. This phenomenon is termed as production carryover in this book. A production carryover is nothing but the production of a product continuously extending to any number of future periods, due to setup of that product in some earlier period, subject to demand and capacity constraints. Researchers have used the term "setup carryover" and "linked lots" interchangeably in the literature, however, this phenomenon is termed "production carryover" in our book. This phenomenon (production carryover of product 1) is indicated in the white region in periods 1, 2, 3 and 4, and is illustrated using Fig. 1.2(i).

In most manufacturing industries, the time to set up the product is smaller compared to the capacity of the period (measured in time units) in which the product is set up. In these industries, when the capacity of a machine has to be utilized efficiently, the idle time present in a period can also be utilized judiciously. Therefore, when there is enough capacity left at the end of a period, it can be utilized in making a complete setup for a product, followed by its production in time period $t + 1$ or the setup started in period t can be split between the periods t and $t + 1$, followed by its production in time period $t + 1$ (this aspect is referred to as setup splitting). Following the inclusion of production carryover, setup splitting has also been considered by researchers in modelling the CLSP in the existing literature. This phenomenon (setup splitting of product 3) is indicated using the shaded region, starting from period 1 and ending in period 2, and is illustrated using Fig. 1.2(ii).

In almost all continuous manufacturing industries such as process industries and some manufacturing industries, due to the presence of long setup times, manufacturers allow the setup to be carried across more than one period (this aspect is called setup crossover). Few researchers attempted to address the CLSP considering setup crossover. This phenomenon (setup crossover of product 5) is indicated using a shaded region which starts in period 1 and ends in period 3, and is illustrated using Fig. 1.2(iii).

In some industries, the setup time of a machine not only depends on the time to set up a product, it also depends upon the product previously setup on the machine. This aspect where the setup time of a product on the machine depends on the time to set up product i' after setting up product i is called sequence-dependent setup. Researchers have also considered the presence of sequence-dependent setup times and setup costs, while addressing CLSP in industries having the presence of sequence-dependent setups. Some researchers have addressed the CLSP with sequence-dependent setups along with production carryover across periods.

Based on the above, literature on capacitated lot sizing problem (CLSP) can be classified into the following categories: (a) CLSP without production carryover across periods and without sequence-dependent setups (Sect. 1.4.1); (b) CLSP without production carryover across periods and with sequence-dependent setups (Sect. 1.4.2); (c) CLSP with production carryover across periods and without sequence-dependent setups (Sect. 1.4.3) and (d) CLSP with production carryover across periods and with sequence-dependent setups (Sect. 1.4.4). The overall classification of literature on CLSP is shown in Fig. 1.3. The classification of CLSP literature based on various categories is listed in Tables 1.1, 1.2, 1.3, 1.4. Section 1.5 provides the summary of this chapter.

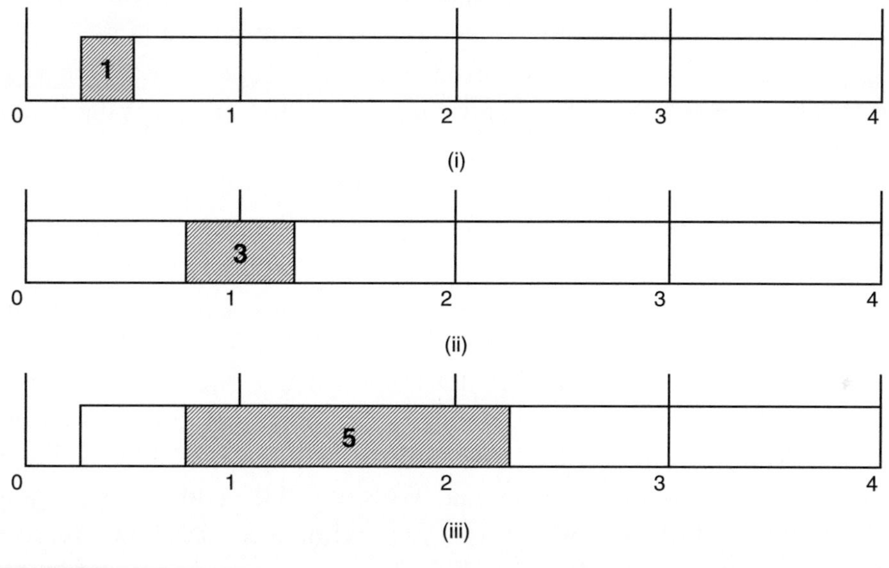

Figure 1.2: Illustration of production carryover, setup splitting and setup crossover phenomena

1.4.1 Literature on CLSP Without Production Carryover Across Periods and Without Sequence Dependent Setups

Initially, authors who worked on CLSP considered only the presence of setup costs (without the consideration of setup times). This problem does not involve the scheduling (timing of production) of products. For each lot produced in a period, only a cost proportional to the setup occurs. Also, the authors assumed the capacity to be uncapacitated or capacitated.

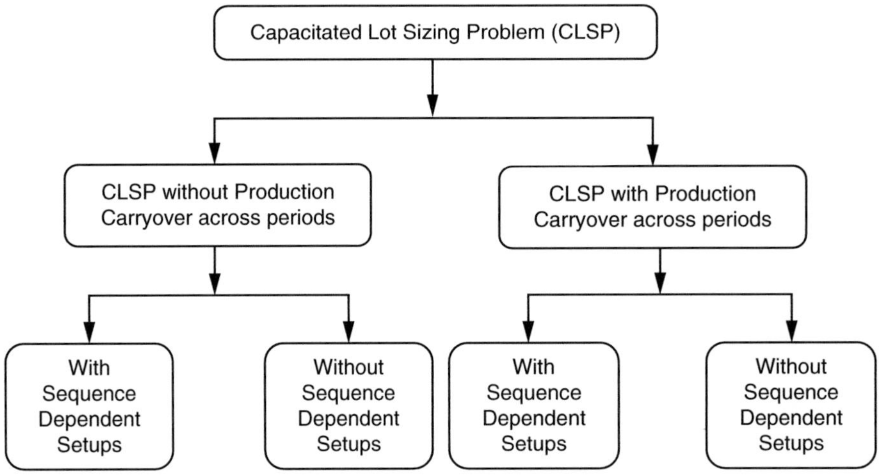

Figure 1.3: Overall classification of the review of literature

Considering the uncapacitated case, in the earliest lot sizing literature, a dynamic programming algorithm for the single-item uncapacitated lot sizing problem was developed by Wagner and Whitin (1958). This algorithm provides an optimal solution to the uncapacitated single-item lot sizing problem. Zangwill (1966) introduced the presence of backorders to the model developed by Wagner and Whitin (1958), and Zangwill (1969) extended the results to include multiple facilities.

Single-resource heuristics also called as "Period-by-period" heuristic (Eisenhut 1975); "Greedy heuristics" by Dogramaci et al. (1981); column generation heuristics (Bahl 1983); and math-programming based heuristics such as "Relaxation heuristics" (Thizy and van Wassenhove 1985) are some of the heuristics developed to solve CLSP with setup cost. The single-item CLSP was proven to be an NP-hard problem by Florian et al. (1980) and Bitran and Yanasse (1982). This CLSP does not involve the presence of setup times. Billington et al. (1983) formulated the standard mathematical formulation for the basic CLSP. Eppen and Martin (1987) solved the multi-item capacitated lot sizing problem using variable redefinition technique. Pochet and Wolsey (1988) addressed the single-machine CLSP with backorders, without setup times and without production carryover, using a shortest-path and an alternative plant location formulation. The shortest-path formulation was solved by standard mixed integer programming procedures, and the alternative plant location formulation was solved using a cutting plane algorithm. The maximum size of problems that can be solved using this algorithm is 100 products and 8 periods. Trigeiro et al. (1989) also formulated the standard mathematical formulation for the basic CLSP.

With positive setup times incorporated in the CLSP, even for finding a feasible solution, the CLSP was proven to be NP-complete by Trigeiro et al. (1989) and Maes et al. (1991). Trigeiro et al. (1989) also developed a Lagrangian based heuristic in order to address the CLSP with setup costs and setup times. *Note:* The math-programming based heuristics such as the Lagrangian heuristic developed by Trigeiro et al. (1989) and Maes et al. (1991) reduce the computational effort by finding the global optimum and get truncated in some way, whereas the "Period-by-period" heuristic (Eisenhut 1975) and the "Greedy heuristic" (Dogramaci et al. 1981) are some heuristics developed using common sense and by understanding the structure of the specific lot-sizing problems in hand. For example, the "Period-by-period heuristic" uses priority rules and determines the production volume in each period separately in a forward or a backward manner, whereas the "Greedy heuristic" starts from an infeasible solution, and by shifting the production volumes to each period it tries to find a feasible solution.

A forward algorithm was proposed by Federgruen and Tzur (1991), to solve the single-item dynamic lot sizing model. Algorithms were proposed based on Wagner and Whitin's dynamic programming algorithm by Wagelmans et al. (1992) and Aggarwal and Park (1993). All the three aforementioned papers aimed at reducing the computational complexity in comparison with the Wagner and Whitin algorithm. The CLSP with parallel resources was addressed by Diaby et al. (1992). Positive setup times were considered in their formulation, without considering production carryover and backorders. Even when number of resources were available for setup, only a single setup was allowed to be performed whenever a product was produced in a period. The problem was solved using a Lagrangian relaxation scheme with sub-gradient optimization which solved instances with up to 5,000 products and 30 periods.

A Lagrangian decomposition procedure was proposed to solve the CLSP by Millar and Yang (1993). Two algorithms for the single-machine lot sizing problem with the presence of backorders were developed by Millar and Yang (1994). One of the algorithms is based on Lagrangian decomposition and the other algorithm is based on Lagrangian relaxation, both neither covering setup times nor production carryover. Tempelmeier and Helber (1994) developed a heuristic for solving the multi-level capacitated lot sizing problem (MLCLSP) for general product structures. Some solution approaches to solve the uncapacitated lot sizing problem were described by Wolsey (1995). Tempelmeier and Derstroff (1996) proposed a Lagrangian based heuristic for solving the multi-level constrained lot sizing problem with setup times. Mixed integer programming (MIP) model formulations were developed by Stadtler (1996) for the dynamic, multi-item MLCLSP. The lot sizing in general assembly systems was addressed by Katok et al. (1998), assuming setup costs, setup times and multiple constrained resources. Stadtler (1997) developed reformulations of the shortest route model for the dynamic, multi-item, MLCLSP, to overcome the major drawbacks in using the original model, even though its LP (Linear Programming) relaxation provided strong bounds. The major drawbacks were as follows: (a) even when standard programming software were used to solve the LP relaxation, problem instances considering 40 products and 16 time periods required a huge amount of computer time to solve the problem, and (b) when large number of time periods were considered, there was a quadratic growth in the number of variables in the model, due to which problems having too many periods could not be solved easily. A modified model was developed by Stadtler (2000), for the dynamic, single-level lot sizing problem which looked at demand forecasts beyond the planning horizon. It was proved that such problems could be solved by exact algorithms such as the Wagner and Whitin (1958) algorithm.

Özdamar and Birbil (1998) addressed the parallel machine lot sizing problem with setup times, without setup costs, with backorders and production carryover. Each machine was allowed to use a certain amount of overtime capacity at respective costs, in addition to the regular capacity. It was assumed that only one of the parallel machines could be used to produce a specific product per period. Three hybrid heuristics, a genetic algorithm (GA) operating in the infeasible region of the solution space, tabu search and simulated annealing (SA) algorithm were developed to solve the problem (by improving the solutions and making them feasible).

The same CLSP was extended by Özdamar and Barbarosoğlu (1999) to multiple production stages, by including setup times and also allowing the presence of backorders. Similar to the above heuristics, a stand-alone simulated annealing procedure was compared with two hybrid heuristics, combining simulated annealing with a genetic algorithm and a Lagrangian relaxation approach, respectively. The Lagrangian relaxation approach provided better results with respect to the solution quality and computational time, for test problems with up to 20 products, 6 periods and 5 machines, on each of the 4 production stages considered.

The capacitated lot sizing problem with overtime decisions and setup times was addressed by Özdamar and Bozyel (2000). They addressed the CLSP, with an objective of minimizing the setup costs and inventory holding costs, thereby determining the quantity of lot sizes to be manufactured over the entire planning horizon. Here, the setup times incurred an additional overtime cost if the overtime capacity was used to set up a product. Since the problem is considered to be an NP-hard problem, various approaches were used to solve the problem. Sub-optimal solutions were provided when the hierarchical production planning approach was used. In the proposed genetic algorithm, an iterative approach was used which omitted the binary variables from the capacity constraints. The proposed simulated annealing algorithm helped in shifting lot sizes between periods,partially or fully. Among the proposed approaches, the simulated annealing and the iterative approach provided the best solutions to the problem considered.

The classical MLCLSP was addressed by Hung and Chien (2000), allowing the presence of setup times and backorders. Three meta-heuristics (tabu search, simulated annealing, and a genetic algorithm) were suggested and analysed by them, since it was very difficult to solve the problem optimally. Integer solutions for the setup variables were found using the meta-heuristics, and the solutions were evaluated by solving an LP problem. Randomly generated test instances with up to 10 periods, 20 items, 5 end-products and 3 stages (considering a single machine in each stage) were considered. The authors observed that tabu search and simulated annealing performed the best among the three heuristics.

Table 1.1: Literature review on CLSP without production carryover across periods and without sequence-dependent setups

Author(s)	Year	Single machine	Parallel/ multiple machines	Single level	Multiple levels	No backorder	Backorders allowed	Production carryover	Setup splitting	Setup carryover	Sequence-dependent setup cost/ setup time
Wagner and Whitin	1958	✓		✓		✓					
Zangwill	1966	✓		✓			✓				
Zangwill	1969		✓	✓			✓				
Eisenhut	1975	✓		✓		✓					
Florian et al.	1980	✓		✓		✓					
Dogramaci et al.	1981	✓		✓		✓					
Bitran and Yanasse	1982	✓		✓		✓					
Bahl et al.	1983	✓		✓		✓					
Billington et al.	1983		✓		✓	✓					
Thizy and Van Wassenhove	1985	✓		✓		✓					
Eppen and Martin	1987	✓	✓	✓		✓					
Pochet and Wolsey	1988	✓		✓			✓				
Trigerio et al.	1989	✓		✓		✓					
Federgruen and Tzur	1991	✓		✓		✓					

(continued)

Table 1.1: (continued)

Author(s)	Year	Single machine	Parallel/ multiple machines	Single level	Multiple levels	No backorder	Backorders allowed	Production carryover	Setup splitting	Setup carryover	Sequence-dependent setup cost/ setup time
Maes et al.	1991	✓		✓		✓					
Diaby et al.	1992		✓	✓		✓					
Wagelmans et al.	1992	✓		✓		✓					
Aggarwal and Park	1993	✓		✓		✓					
Millar and Yang	1993	✓		✓		✓					
Millar and Yang	1994	✓	✓	✓			✓				
Tempelmeier and Helber	1994		✓		✓	✓					
Wolsey	1995	✓		✓		✓					
Stadtler	1996		✓		✓	✓					
Tempelmeier and Derstroff	1996		✓		✓	✓					
Stadtler	1996	✓			✓	✓					
Stadtler	1997	✓			✓	✓					
Katok et al.	1998		✓		✓	✓					
Özdamar and Birbil	1998		✓	✓		✓					
Özdamar and Barbarosoglu	1999		✓		✓		✓				
Hung and Chien	2000		✓		✓		✓				
Stadtler	2000	✓		✓		✓					
Özdamar and Boyzel	2000	✓		✓		✓					

Note: In Tables 1.1–1.4, a machine is also treated as a resource or a facility in some of the papers which have been reviewed. Also, the level of product can also refer to the stage of a product (i.e., single stage or multiple stages) in some of the papers. In all the tables, literature on capacitated lot sizing problem (CLSP) has been listed, except in Table 2.1 where some of the earliest papers in lot sizing deal with uncapacitated lot sizing problems. They are, papers by Wagner and Whitin (1958), Zangwill (1966), Zangwill (1969), Federgruen and Tzur (1991), Wagelmans et al. (1992), and Aggarwal and Park (1993)

1.4.2 Literature on CLSP Without Production Carryover Across Periods and with Sequence Dependent Setups

A single-machine MIP model with sequence-dependent setup times and backorders was addressed by Smith-Daniels and Smith-Daniels (1986). The model was developed using a combination of the small bucket and the big bucket approaches. Several product families consisting of various items were considered, and in each period, only one product family could be produced (small-bucket model). While sequence-dependent setup times were considered between items, setup times were not considered between families as family setups take place between periods. Also, sequence-independent family setup costs were considered, but item-to-item setup costs were not considered. Using standard procedures, a simplified version of the model including four items and five periods was solved. Grünert (1998) focused on the CLSP for a multi-level product structure considering sequence-dependent setup costs and setup times, and developed a solution algorithm based on Lagrangian decomposition and tabu search.

Kang et al. (1999) developed a model for lot sizing and scheduling in a multiple machine environment considering sequence-dependent setup costs and setup times. An optimization approach based on column generation/branch & bound was developed. Using five real-world problem instances gathered from the industry, the proposed approach was heuristically adapted and tested. Clark and Clark (2000) developed a model for the multi-product lot sizing problem with sequence-dependent setup times. Since the model was too large to be solved optimally, approximate models were developed, and using a rolling horizon basis, exact schedules for the next immediate period were only obtained. For problems of large size, in order to overcome the enormous computing time incurred by approximate models, a rapid solution procedure was developed.

Timpe (2002) proposed an algorithm combining MIP and constraint programming to solve the CLSD (capacitated lot sizing problem with sequence dependent setups). Schedules within periods were derived using constraint programming, and an MIP was used for determining the lot sizes of the products. The problem was taken from an industry, and the maximal and minimal production quantities were determined for each lot and for each period independently. Lang and Shen (2011) developed an MIP model for the single-level dynamic capacitated lot sizing and scheduling problem with sequence-dependent setup costs and setup times incorporating options for product substitution. The development of the model was motivated from an industry manufacturing plastic sheets which are used as an inter-layer in car windshields. Several variants of these sheets which vary in length, colour, material, nature of surface, width and thickness were manufactured in the machine, and sequence-dependent setups occurred when changing from one foil variant to another. Apart from the mathematical model, a heuristic based on the fix-and-optimize approach of Sahling et al. (2009) was adapted and combined with the time oriented decomposition heuristic to solve this problem.

Table 1.2: Literature review on CLSP without production carryover across periods and with sequence-dependent setups

Author(s)	Year	Single machine	Parallel/ multiple machines	Single level	Multiple levels	No backorder	Backorders allowed	Production carryover	Setup splitting	Setup carryover	Sequence-dependent setup cost/ setup time
Smith-Daniels and Smith-Daniels	1986	✓		✓			✓				✓
Grunert	1998	✓			✓	✓					✓
Kang et al.	1999		✓	✓		✓					✓
Clark and Clark	2000		✓	✓			✓				✓
Timpe	2002		✓	✓			✓				✓
Lang and Shen	2011	✓		✓			✓				✓

1.4.3 Literature on CLSP with Production Carryover Across Periods and Without Sequence Dependent Setups

Lambrecht and Vanderveken (1979) were the first to present a mathematical model for the CLSP with production carryover across periods. However, the linking constraints (production carryover constraints) were dropped from the model, and the model was solved as a basic CLSP because the problem was considered too difficult to be solved during that time. The CLSP with production carryover across periods and without the presence of backorders was addressed by Dillenberger et al. (1993) (for parallel machines). A practical production planning problem in IBM was the motivation for this research. In this planning problem, production was allowed to take place only in the periods in which the demand was present. The production could be backordered only when there was insufficient capacity to meet the demand (backordering is different from backlogging in such a way that the backordered units can be produced in some future period instead of meeting the demand in the current period, whereas backlogged units cannot be produced in the future period and the demand is lost). The authors developed a new mathematical model (MIP) based on which a solution algorithm known as the fix-and-relax heuristic approach was developed. The fix-and-relax algorithm consisted of branch & bound scheme in which the binary variables to be branched were ordered based on the sequence of the period.

Dillenberger et al. (1993)'s model was later extended by Dillenberger et al. (1994) by incorporating different kinds of resources like energy or components. Storable resources were also considered, where their capacity could be consumed in later periods if not consumed in the current time period. Product families were also considered. Here, a major setup was allowed to take place when switching of products occurred from one family to another family, and a minor setup was allowed to take place while switching occurred between products within the same family, leading to partial sequence-dependent setups. Setup times and setup costs were considered for both major and minor setups. A partial branch & bound (B & B) scheme that iterates on a period-by-period basis was developed as the solution procedure to the problem. In every iteration of the B & B scheme, setup variables relating to the actual period were fixed to binary values, and the binary restrictions for later periods were relaxed. Practical problems up to 40 products, 6 periods and 15 machines were solved using this procedure.

Haase and Drexl (1994) introduced the name CLSPL (capacitated lot sizing problem with linked lot sizes) for the capacitated lot sizing problem with production carryover, and developed a mathematical model for the problem (considering a single machine, and the presence of backorders was not considered). The problem was solved using a stochastic heuristic which moved backwards, from the last period to the first period thereby scheduling setups and determining the production volume. The heuristic used a randomized regret measure to make lot sizing decisions. Gopalakrishnan et al. (1995) developed a mixed integer linear programming (MILP) model for the CLSP with production carryover. The problem was motivated by a large paper manufacturer who produced many paper products. Each family of paper product consisted of various types of products. A major setup cost occurred when a product from another family had to be produced after the producing a product from the current family. When another product within the same family had to be produced, the setup cost incurred was insignificant (due to a minor setup). In a minor setup, the setup cost and setup time were assumed to be same across all products. An MILP model was developed based on this, in order to reduce the setup, holding and fixed cost of production of all products. The setup time was assumed to be same across all products of all families and backlogs were not considered.

Later, the model was extended to include multiple identical machines in which multiple families of the products were produced. For products belonging to the same family, there was a special tool needed which remained fixed to the machine. Hence, the decision problem was to determine the number of tools required to produce a particular type of product, and their assignment to the machines, along with the lot sizing and partial sequencing decisions for the products belonging to various families. Hence, the tool carrying cost was also included in the objective function.

Haase (1998) formulated a model for the CLSP with lot sizes linked to the adjacent periods (production carryover across periods). The setup state of a product could not be preserved more than the boundary of one period. To solve this problem and compute the solution in a reasonable time, a priority rule based heuristic was developed which is a backward oriented scheduling procedure. A simple local search method was applied to obtain low cost solutions. Sox and Gao (1999) developed a mathematical model for the CLSP with production carryover across periods. Multiple products were allowed to be produced on a single machine, thereby minimizing the setup cost and inventory holding cost which are time dependent. The presence of backorders was not allowed. Efficient reformulations were developed to find optimal solutions. A Lagrangian decomposition heuristic based on sub-gradient optimization and dynamic programming was also developed in order to find near optimal solutions for the problem. A simple forward period-by-period heuristic was proposed by Gao (2000) for the CLSP with production carryover across periods, with no backorders and no setup times. The proposed heuristic delivered quick solutions with higher cost.

Gopalakrishnan (2000) extended his previous work of modeling the CLSP with production carryover with constant setup time, by including product-dependent setup times and costs. An MILP model was developed for this problem. Gopalakrishnan et al. (2001) developed a tabu search heuristic for the capacitated lot sizing problem with production carryover. The heuristic was developed based on fives moves, i.e., (a) Swap items within a period (SWAP); (b) Move setup at the beginning of a period to an earlier period (SETUP1); (c) Move end-of-period setup to the following period (SETUP2); (d) Move production lot to an earlier period (LOT1) and (e) Move production lot to a later period (LOT2), for sequencing as well as lot sizing decisions. The efficiency of the developed heuristic was tested using problem instances of Trigeiro et al. (1989), and the computational times were also compared.

Porkka et al. (2003) developed a production carryover model using setup times. The study also involved in comparing three models, a CLSP NCO (non-carry over) model, a CO (carry over) model, and a CCO (compressing carry over) model which allowed production stoppages between successive production of the same product, without setting up the product twice. Heuristics were also developed for the NCO model to allow carryovers in between (post-optimal modification of the carryover) and reduce the cost of setup. Quadt and Kuhn (2003) solved the CLSP with production carryover across periods using a heuristic solution algorithm. The algorithm was based on an MIP formulation developed for parallel resources and allowed for the presence of backorders. The solution approach was applied to a semi-conductor industry (consisting of a huge number of parallel resources). Since a huge number of resources were present in that industry, the resources were represented using integer variables which held a certain setup state, instead of using binary variables in the same place. The heuristic gave good solutions for problems of larger sizes compared to the developed MIP model.

Suerie and Stadtler (2003) proposed a new model for the capacitated lot sizing problem with production carryover across periods. The proposed model was incorporated into a time-oriented decomposition heuristic. The solution approach was based on an extended model formulation and valid inequalities. Through extensive computational tests it was proven that the proposed approach was superior when compared to the then so far published solution algorithms. Briskorn (2006) wrote a note on the Lagrangian relaxation technique developed by Sox and Gao (1999), for the capacitated lot sizing problem using production carryover. The author provided an explanation to show that the algorithm did not provide optimal solutions to the sub-problem considered. The author also corrected the flaws in order to solve the sub-problems optimally.

Karimi et al. (2006) developed a mathematical model for the capacitated lot sizing problem with production carryover, and included the presence of backorders in the model. A tabu search heuristic was also proposed to solve this problem. In order to provide an improved solution as the starting solution to the tabu search heuristic, a heuristic for obtaining a feasible solution followed by the determination of an improved feasible solution was also proposed before solving the tabu search heuristic. The heuristic proposed to find an initial feasible solution, and consisted of four elements, namely, (a) rule for demand shifting; (b) rule for determining lot sizes; (c) checking feasibility conditions and (d) determination of production carryover. The initial feasible solution was improved by adopting the corresponding setup and setup carryover schedule, and it was re-optimized by solving a minimal-cost network flow problem.

Nascimento and Toledo (2008) proposed a mathematical formulation for the multi-plant capacitated lot sizing problem with production carry over. A heuristic was proposed based on GRASP (greedy randomized adaptive search procedure) along with a path-re-linking procedure to solve the problem. The results of the heuristic were compared using the gap between the original heuristics based on GRASP with path-re-linking and without production carryover (proposed by Nascimento and Toledo (2008)), and it was seen that there was a very low increase in computational time compared to the one without production carryover. Caserta et al. (2009) presented a hybrid algorithm for the MLCLSP with production carryover. In the proposed hybrid approach, a meta-heuristic, an LP solver and a dynamic programming scheme was iteratively used within the Lagrangian framework. In order to find feasible solutions in the neighbourhood of an incumbent infeasible solution, three repair heuristics were also presented. Using 480 benchmark instances, the algorithm was tested and the results obtained were good in terms of quality and computational time.

Quadt and Kuhn (2009) addressed the capacitated lot sizing problem with production carryover across periods for a semiconductor assembly facility which allowed the presence of backorders. A mixed integer programming model and a solution procedure were proposed to address this problem. The solution procedure which was developed based on a novel aggregate model used integer instead of binary variables. The model was embedded in a period-by-period heuristic, and was solved to optimality using CPLEX. Using a subsequent scheduling routine, the products were loaded and sequenced on parallel machines. Six variants of the heuristic were presented in this paper. Sahling et al. (2009) developed a model for the capacitated lot sizing problem for products having multiple levels like a general Bill-of-Materials (BoM) structure. The BoM structure has a minimum lead time of one period, and the model also incorporates production carryover across periods. A new algorithm was developed which solved a series of MIP models iteratively in a fix-and-optimize approach, in which under each iteration, a small subset of the variables were fixed and a subset of the variables were optimized.

Tempelmeier and Buschkühl (2009) developed a new model for the dynamic MLCLSP with production carryover across periods. The products had a BoM structure with several end products, and each with a dynamic external period demand over a finite time horizon. All the products were assumed to be manufactured on a single resource/machine with a finite capacity. The presence of backorders was not allowed. Apart from the mathematical model, a Lagrangian based heuristic was also developed, and the heuristic was tested on invented data as well as industrial data. The solutions obtained were of good quality. Caserta et al. (2010) presented a math heuristic algorithm for the MLCLSP with production carryover across periods. Mathematical programming techniques were used in the heuristic algorithm in a meta-heuristic fashion, thereby solving smaller portions of the original problem iteratively.

Oztürk and Ornek (2010) developed a mathematical model for the capacitated lot sizing problem with linked lot sizes (i.e., production carryover across periods), such that multiple products having multiple levels could be manufactured on multiple resources. In the product structure, all items had dependent demand as well as independent demand. Backorders were allowed for the independent demands only and not the dependent demands as it may affect the whole demand balance of the product structure. Computational results were taken for this model in a job shop environment, and the results were presented along with a numerical example. Wu and Shi (2011) presented theoretical results on four formulations, for the capacitated lot sizing problem with production carryover across periods (mentioned as CLSPL in that paper), where LP relaxations of the formulations were carried out and only a subset of binary variables were relaxed to be continuous. The four formulations were the facility location (FL) formulation proposed by Bilde and Krarup (1977), shortest route (SR) formulation proposed by Eppen and Martin (1987), simple multi-commodity (SMC) formulation proposed by Rardin and Wolsey (1993), and the strengthened shortest route (SSR) formulation proposed in their paper. In their paper, it was proved that the FL and SR formulation were equivalent and the SMC and the SSR formulation were equivalent. The formulations were proved to be equivalent in order to choose the best formulation, to apply certain decomposition based heuristics so as to find the upper bound (which would be the same for equivalent models). However, the differences were with respect to the solution time taken for solving them. Computational experiments were conducted by using the relax-and-fix method using an example. The FL formulation proved to be the most efficient in terms of LP relaxation, and the relax-and-fix approach proved to be the most efficient for medium sized problems. The SR formulation proved to be the most efficient for problems of large sizes.

Goren et al. (2012) proposed a novel hybrid approach by combining two approaches for solving the capacitated lot sizing problem with production carryover across periods. The two approaches were namely the genetic algorithm and a fix-and-optimize heuristic. In order to reduce the solution space and provide a feasible solution, a new initialization scheme was also proposed. Using some benchmark problem instances a comparative experimental study was carried out, and it was seen that due to hybridization of the genetic algorithm with the fix-and-optimize heuristic, the performance improved compared to using a pure genetic algorithm. Caserta and Voß (2013) presented a meta-heuristic black box framework to tackle the MIP problems through a "constrained black box approach". The approach used an MIP solver to explore some portions of the solution space, to optimality. The meta-heuristic governed the selection of portions of the solution space, so that the likelihood for finding the improved solutions was high by keeping a low time to search the solution space. The meta-heuristic black box framework was tested on the capacitated lot sizing problem with production carryover, and it was proved that the proposed method was competitive with ad-hoc heuristics, and outperformed the black box solver when directly used on the original problem. Wu et al. (2013) developed an MIP formulation for the capacitated lot sizing problem, consisting products having multiple levels, and assuming the presence of backorders and setup carryover. Based on the formulation, a time-oriented decomposition heuristic was proposed in which the improvement and construction heuristics were effectively combined in order to overcome the weaknesses of the classical time-oriented decomposition algorithm.

CLSP with Production Carryover Across Periods and Setup Splitting Between Two Periods

Mohan et al. (2012) developed an MIP model for the capacitated lot sizing problem with production carryover across periods and setup splitting between two periods. The developed model with setup splitting was compared with some earlier model without setup splitting, and was found to give better solutions. Ramya et al. (2016) also addressed the same problem with the help of mathematical models.

Table 1.3: Literature review on CLSP with production carryover across periods and without sequence-dependent setups

Author(s)	Year	Single machine	Parallel/ multiple machines	Single level	Multiple levels	No backorder	Backorders allowed	Production carryover	Setup splitting	Setup carryover	Sequence-dependent setup cost/ setup time
Lambrecht and Vanderveken	1979		✓	✓			✓	✓			
Dillenberger et al.	1993		✓	✓			✓	✓			
Dillenberger et al.	1994		✓	✓			✓	✓			
Haase and Drexl	1994	✓		✓		✓		✓			
Gopalakrishnan et al.	1995	✓	✓	✓		✓		✓			
Haase	1998	✓		✓		✓		✓			
Sox and Gao	1999	✓		✓		✓		✓			
Gao	2000	✓		✓		✓		✓			
Gopalakrishnnan	2000	✓		✓		✓		✓			
Gopalakrishnnan et al.	2001	✓		✓		✓		✓			
Porkka et al.	2003	✓		✓		✓		✓			
Quadt and Kuhn	2003		✓	✓			✓	✓			
Suerie and Stadtler	2003		✓	✓	✓	✓		✓			
Briskorn	2006	✓		✓		✓		✓			
Karimi et al.	2006	✓		✓			✓	✓			

Author(s)	Year	Single machine	Parallel/ multiple machines	Single level	Multiple levels	No backorder	Backorders allowed	Production carryover	Setup splitting	Setup carryover	Sequence-dependent setup cost/ setup time
Nascimento and Toledo	2008		✓	✓		✓		✓			
Sung and Maravelias	2008	✓		✓			✓	✓		✓	
Caserta et al.	2009		✓		✓	✓		✓			
Quadt and Kuhn	2009		✓	✓			✓	✓			
Sahling et al.	2009		✓		✓	✓		✓			
Tempelmeier and Buschkuhl	2009		✓		✓	✓		✓			
Caserta et al.	2010		✓		✓	✓		✓			
Oztürk and Ornek	2010		✓		✓		✓	✓			
Wu and Shi	2011		✓		✓	✓		✓			
Goren et al.	2012	✓		✓		✓		✓			
Mohan et al.	2012	✓		✓		✓		✓	✓		
Belo-Filho et al.	2013	✓		✓			✓	✓		✓	
Caserta and Voß	2013		✓		✓	✓		✓			
Wu et al.	2013	✓		✓		✓		✓			

CLSP with Production Carryover Across Periods and Setup Crossover Across Many Periods

Sung and Maravelias (2008) developed a new model for the capacitated lot sizing problem with production carryover across periods. The authors addressed the case of long setup times by allowing the setup to crossover across more than one period. The formulation was extended to include the presence of idle time, parallel machines, periods of non-uniform length, backorders, lost sales and several product families.

Belo-Filho et al. (2013) developed two mathematical models for the capacitated lot sizing problem with production carryover and setup crossover across periods. Setup crossover is essential for products having long setup times. The presence of backorders was permitted. Both the developed formulations were compared with a previous state-of-the-art formulation by Sung and Maravelias (2008).

1.4.4 Literature on CLSP with Production Carryover Across Periods and with Sequence Dependent Setups

Smith-Daniels and Ritzman (1988) developed a multi-level MIP model formulation considering sequence-dependent setup times for the production planning problem in a food industry (it is an extension of the problem addressed by Smith-Daniels and Smith-Daniels (1986) to a multi-level case). Product families were considered and more than one family could be produced per period (big-bucket model). Although sequence-dependent setup times were present when family changeovers occurred, item-to-item setup times were not present. Setup costs were not considered in the modelling. Families within a period were automatically sequenced when a family setup was carried over from one period to another. Using standard procedures, a single instance consisting of three families with two items each, three periods and two stages was solved. A single-machine lot sizing and scheduling problem with production carryover and sequence-dependent setups in a chemical manufacturing environment was addressed by Selen and Heuts (1990). Here the production lots constituted an integer number of pre-defined quantities or batches, the inventory capacity was assumed to be limited, and the presence of backorders was not allowed. A lot-for-lot based solution procedure was proposed which started with a feasible production plan, and shifted complete production lots to earlier periods with the target period moving forward in a period-by-period manner. A sample problem instance with 20 products and 10 periods was also illustrated in the paper.

Haase (1996) developed a mathematical model for the capacitated lot sizing problem considering sequence-dependent setup costs, with an objective of minimizing the setup and holding costs, and without considering the presence of backorders. The setup state of a product was allowed to be carried from period to period in the presence of idle time. A single-stage system was considered in which a number of items had to be produced on a single machine. Apart from the mathematical model, a heuristic which applied priority rules was also considered in order to obtain the solution to the problem. Haase and Kimms (2000) dealt with the lot sizing and scheduling problem for a single-machine production system, considering items having a single stage, and assuming sequence-dependent setup costs and setup times. Production was allowed to be carried over across periods. The problem was inspired from a practical case at Linotype-Hell AG, Kiel (Germany), manufacturing high technology machines for the setting and printing business. Here a numerically controlled milling machine was identified as being the bottleneck machine because setting up of the milling machine required mounting specific tools and holders. The sequence-dependent setup time consists of taking off the tools and holders of the previous setup, loading another program, and again mounting tools and holders for the next setup. A large bucket MIP model was developed. Rescheduling was also integrated into the model which helped to accommodate further changes in the demand matrix when new information from the customers was received. A fast enumeration scheme was proposed which used the B & B. A "zero-inventory" property was assumed such that the setup operations for a product could occur only in those periods in which the inventory for that product was zero in the previous period (which generally is not considered in the CLSP).

Almada-Lobo et al. (2007) developed two MIP models for the production planning problem in a glass container industry. Sequence-dependent setup costs and setup times are prevalent in these industries when there is a switch over from one product to another. The glass container industry has a semi-continuous manufacturing process, the glass production being the continuous part and the container production being the discrete part. Molten glass is fed into the moulding machine continuously from the furnace, and when the product changeover happens, the gobs are discarded and remelted again in the furnace which results in the waste of energy. In order to incorporate fewer and faster product changes, sequence-dependent setup times and costs are considered in this industry. The models developed incorporate production carryover across periods. Apart from the mathematical models, a five-step heuristic was also developed for finding feasible solutions. Kovács et al. (2009) addressed the capacitated lot sizing and scheduling problem with sequence-dependent setup times and setup costs for a single machine, by proposing a new MIP model. Production could be carried over across periods. The parameters for this problem were generated by a heuristic procedure in order to establish a tight formulation, and the model helped to solve large size problems.

Almada-Lobo and James (2010) addressed the capacitated lot sizing and scheduling problem with sequence-dependent setup times and setup costs, considering the possibility of occurrence of production carryover across periods. The exact method developed failed to solve medium size problem instances, CLSP being an NP-hard problem. Hence, the problem was solved using a tabu search and variable neighbourhood search meta-heuristic. The performance of these heuristics was compared over time. The effectiveness and efficiency of the proposed approaches were shown by solving medium to large size problem instances. Mohammadi et al. (2010) developed an exact model for the capacitated lot sizing and scheduling problem with sequence-dependent setups considering multiple products with multiple levels, and production carryover. MIP-based heuristics were developed to solve large instances of the problem. Apart from this, two lower bounds were developed to test the accuracy of the heuristics, and were also compared against the optimal solution of the problem.

James and Almada-Lobo (2011) solved the single- and parallel-machine capacitated lot sizing and scheduling problem assuming sequence-dependent setup costs and setup times using a general purpose heuristic approach, i.e., by combining meta-heuristics and mixed integer programming, in order to find high quality solutions. In the proposed approach, the setup state of a product was allowed to be carried from one period to another. CLSP being an NP-hard problem, solving medium sized problems was found to be hard using exact methods (when solved using commercial solvers). Therefore, based on the MIP formulation, a construction, an improvement and a search heuristic was developed, and the performance of the proposed heuristics was compared with the existing heuristics in literature. The effectiveness and efficiency of the proposed approaches were tested using a set of computational experiments.

Kwak and Jeong (2011) considered a special structure of sequence-dependent setups for the capacitated lot sizing and scheduling problem for a single machine, assuming production carryover across periods. This CLSP with sequence-dependent setup times was solved using a two-level hierarchical planning procedure, where the sequence of production of the products along with the production quantity was determined, thereby minimizing the sequence-dependent setup costs and holding costs for multiple products. A special structure of sequence-dependent setups was considered when larger products were setup on the machine (the setup times for larger products were considered to be very high). Shim et al. (2011) considered the capacitated lot sizing problem for a single machine with the setup state preservation across periods, thereby determining the lot sizes and the sequence of lots, while satisfying the demand requirements and the capacity of the machine in each period of the planning horizon. The problem was motivated by an industry manufacturing corrugated cardboards from paper wastes. It is a process industry where the setup times are dependent on the sequence of the lots produced. This problem was addressed with the objective of minimizing the inventory and setup costs over the planning horizon. A two-stage heuristic was developed owing to the complexity of the problem which used various priority rules to select the items to be moved, and a forward and backward improvement method was also proposed. Clark et al. (2014) addressed the lot sizing and scheduling problem with sequence-dependent setups, and assumed that the setup times were not triangular (i.e., it can be optimal in some situations to produce more than one lot in a period for a product). A mathematical model was proposed for this problem assuming production carryover across periods. However, the carrying of setup states across two periods (i.e. allowing a setup to begin and end in two different periods due to long setup times), which are present in most process industries and complex machine-tooling changeovers, was not assumed. Guimarães et al. (2014) developed a new formulation for the lot sizing and scheduling problem with sequence-dependent setups using commodity flow based sub-tour elimination constraints. In their work they assumed production carryover across periods but setup crossover was not allowed (i.e. allowing the setup to start in one period and end in some later period).

Table 1.4: Literature review on CLSP with production carryover across periods and with sequence-dependent setups

Author(s)	Year	Single machine	Parallel/ multiple machines	Single level	Multiple levels	No backorder	Backorders allowed	Production carryover	Setup splitting	Setup carryover	Sequence-dependent setup cost/ setup time
Smith Daniels and Ritzman	1988	✓			✓		✓	✓			✓
Selen and Heuts	1990	✓		✓		✓		✓			✓
Haase	1996	✓		✓		✓		✓			✓
Haase and Kimms	2000	✓		✓		✓		✓			✓
Almada-Lobo et al.	2007	✓		✓		✓		✓			✓
Kovács et al.	2009	✓		✓		✓		✓			✓
Almada-Lobo and James	2010	✓		✓		✓		✓			✓
Mohammadi et al.	2010		✓		✓	✓		✓			✓
James and Almada-Lobo	2011	✓	✓	✓		✓		✓			✓
Kwak and Jeong	2011	✓		✓			✓	✓			✓
Shim et al.	2011	✓				✓		✓			✓
Clark et al.	2014	✓		✓			✓	✓			✓
Guimarães et al.	2014	✓		✓		✓		✓			✓

1.5 Integrated Decision Making in Supply Chains

Supply chain management involves long term planning (strategic decisions), medium term planning (tactical decisions) and short term planning (operational decisions). Long term decisions which are made in a supply chain involve location decisions of manufacturing plant/distribution center and investing on highly expensive technologies and machines. Medium term decisions which are made in a supply chain involve the selection of suppliers, deciding production capacities, forecasting demand, planning storage and material handling. Some operational decisions include scheduling of production, distribution, ordering and shipping in a supply chain, inventory management and order management. A short term scheduling model helps in determining the jobs that are scheduled inside the facility associated with parameters such as setup costs/setup time (sequence dependent/ independent), and meeting an objective. Many decisions can be made such as determining the lot size of production of each product, number of early or late shipments, etc. A medium term production planning model optimizes several stages of a supply chain, where each stage has one or more facilities, and the model is designed to determine the production in these facilities in each time period taking into account the associated costs and objectives. Mostly, scheduling in the context of supply chain focuses more on a single stage in a supply chain and optimizes over a short term horizon. Scheduling uses detailed information, whereas medium term supply chain planning covers multiple stages and uses the aggregated information. Scheduling is an integral part of enterprise planning in supply chains (Kreipl and Pinedo 2004).

Supply chain management consists of two major components, namely, network organization and inter-organizational collaboration. Information and communication technology helps to co-ordinate among various stages in a supply chain which are addressed in advanced planning systems (APS) (Stadtler 2005). Integration between different units of an organization is very important to improve the competitiveness of the supply chain. Information technology helps in coordinating the material, financial and information flows throughout the supply chain. Decision making needs to be performed simultaneously among all entities of a supply chain at the operational level.

APS has also become a leading edge in the context of enterprise planning because manufacturers have to deal with finite capacity scheduling and have to follow constraint-based planning. Since timing and reliable delivery of a product has become an important factor in this competitive market, meeting due dates has become very important for survival. When finite capacity is there and due dates are hard to be met, outsourcing to subcontracts can also be done with the help of advanced planning. The use of alternative machines has also been considered when there are more jobs to be scheduled. APS is helpful by making use of mathematical models and meta-heuristics. The APS supply chain planning matrix considers: (1) Long term—strategic network planning; (2) Mid-term—master planning and demand planning and (3) Short term—material requirements planning, production planning, transport planning and demand fulfilment. In the production planning and scheduling the levels of planning detail involve shifts, machine groups, flow lines, etc. The production planning which is a part of the system helps to communicate to the other parts of the supply chain to plan their operations accordingly. Based on the capacity constraints, the suppliers can be provided with advance shipment notices, and the dealers/retailer can stock more based on production constraints (Stadtler and Kilger 2002).

A close coordination between finance, marketing, production, procurement and logistics is important in firms. Therefore, members within the supply chain can be informed via EDI (electronic data interchange) instantaneously and cheaply about sudden break in production line, new incoming orders, increase in capacity, etc. In a consumer goods manufacturing industry, supply chain planning elements cannot operate independently. For example, distribution planning cannot exist without a production planning system. Though scheduling comprises the short term plan of the supply chain system, the lots are scheduled according to the due dates of the customers who form the down-stream of a supply chain. Updated demand data also flows into the production environment to adjust production schedules (lot sizing in a rolling time horizon). Interruption and delays are common in a complex production environment and scheduling helps to capture various parameters existing in a supply chain. Electronic goods industries, such as computer assembly supply chain, are perhaps less capacity-constrained, but more material constrained supply chain. Therefore, scheduling of production can be planned only by knowing the availability of the components and lead-time of manufacturing of components. Manufacturing execution systems (MES) have been developed to easily configure real time control devices at the shop floor. APS can be linked with real time control devices for detailed scheduling similar to the MES (Stadtler 2005).

1.6 Summary

The history of the production planning problem and the developments in lot sizing literature show that lot sizing is a vastly researched area. Various characteristics and attributes of the lot sizing problems lead to the complexity of modelling these problems. The classification of several lot sizing problems into continuous and dynamic lot sizing problems (consisting of small bucket and big bucket lot sizing problems) has been presented in this chapter. Different classes of lot sizing problems exist in literature such as ELSP, DLSP, CSLP, PLSP and CLSP. The capacitated lot sizing problem (CLSP) which is the focus of this chapter is a large bucket lot sizing model since it can incorporate the production of multiple products within a period. Various authors have contributed to the CLSP literature starting from the inclusion of setup costs and setup times in the model formulation, followed by the development of heuristics for the basic CLSP problem. Later, contribution to the CLSP literature involved the introduction of the concept-production carryover across periods. Recent literature on CLSP involves the effective capacity utilization, by developing models for the CLSP problem with production carryover and setup splitting, and the latest literature involves in developing exact methods to address the same problem in the context of process industries which have the presence of long setup times (i.e., allowing production carryover and setup crossover across periods). The phenomenon where the setup state of a product is carried from one period to another in order to avoid multiple setups for the same product in consecutive periods is termed setup carryover in literature, whereas it is called a production carryover in this book. Also, the phenomenon where the setup of a product having long setup time is carried over across periods is called setup crossover in this book as well as in the literature. In the next chapter a motivation to study the lot sizing in continuous manufacturing (process industries) is presented.

CHAPTER 2

CLSP: Real Life Applications and Motivation to Study Lot Sizing Problems in Process Industries

2.1 Production Planning in Discrete Manufacturing Industries and Continuous Manufacturing Industries

Generally, the CLSP addresses the production planning problem in discrete manufacturing industries and continuous manufacturing industries. A brief explanation about the production planning in discrete manufacturing industries and continuous manufacturing industries with relevant examples is presented in Sects. 2.1.1 and 2.1.2, respectively.

2.1.1 Discrete Manufacturing Industries

Discrete manufacturing industries are those industries where the machine can be stopped at any time and started at any time. The machine can also remain idle during the course of production of the product as it will not affect the physical or chemical properties of the product. Some examples of discrete manufacturing industries are automobile assembly industry, industry manufacturing automobile parts, furniture industry, manufacturer of smart phones, etc. Gnoni et al. (2003) dealt with a lot sizing and scheduling problem in a multi-site manufacturing industry, manufacturing braking equipment for the automobile industry. The authors claim that sequence-dependent setups are also present in such discrete manufacturing industries; i.e., the production sites have critical bottleneck resources which cause significant sequence-dependent setup times. Breakdown and repair of those machines also contribute to this effect. Haase and Kimms (2000) address the lot sizing and scheduling problem in a company manufacturing high technology machines. Here, sequence-dependent setup time occurred in the numerically controlled milling machine, when the tools and holders of the previous setup were removed followed by loading of another program, and the tools and holders were again mounted for the next setup.

2.1.2 Continuous Manufacturing Industries

Continuous manufacturing industries are distinct, and they are different from discrete manufacturing industries. Continuous manufacturing is present in almost all process industries and in some manufacturing industries whose end products may be discrete, whereas the process to manufacture them is continuous. The American Production and Inventory Control Society (APICS) defines process industries as "Businesses that add value by mixing, separating,

R. Ramya et al., *Capacitated Lot Sizing Problems in Process Industries*, https://doi.org/10.1007/978-3-030-01222-9_2

forming, or chemical reactions. Processes may be either continuous or batch, and usually require rigid process control and high capital investments". The definition states that firms involving production of processed food, paper, cardboard, chemicals, crude oil, rubber and plastic goods, synthetic threads and fibres, building materials such as cement, pottery and glass belong to the category of process industries. Fransoo (1993) classified process industries as batch process industries and flow process industries.

Batch Process Industries

The general characteristics of these industries are: (a) they have several processing steps; (b) product routings are different; (c) material flow is of assembly type and (d) the final product is of high added value. Examples of batch process industries are pharmaceutical industry, chemical manufacturing industry, die casting industry and rubber/plastic manufacturing industries.

Die Casting Industry

In die casting industries, the production starts immediately after the completion of setup of a product. Heating the furnace of the machine up to the right temperature is the process of setup for casting. Once the furnace is heated, the metal (such as aluminium or zinc) is melted and the molten metal is passed through the mould/die under high pressure to produce the castings, in order to obtain the proper shape of the final product. If there is a break in between the heating of the furnace for melting the metal, and the production of the castings, the metal starts getting solidified even before it passes through the casting mould/die. This shows that continuous manufacturing should take place in order to obtain proper characteristics of the final product. The process of die casting is shown in Fig. 2.1.

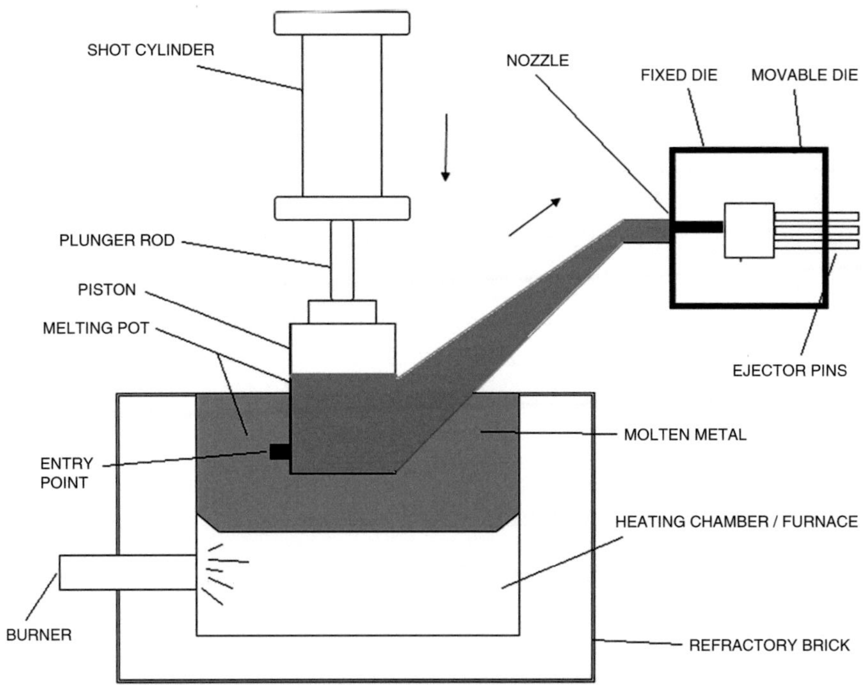

Figure 2.1: Die casting process: A schematic diagram

Rubber Accessories Manufacturing Industry

The injection moulding machine is used to manufacture rubber products which are used in automobiles, and for various other purposes. Some of the rubber parts made with the use of injection moulding machines are balls, gears, plastic caps, etc. In the injection moulding machine, heating the injection unit of the injection moulding machine up to the right temperature is the process of setup. The raw material (plastic raw material granules) is heated in the injection unit until the molten state, and the injection of the molten material into the mould is done immediately so that the production process starts thereafter. Therefore, in this industry, production starts immediately after setup completion, and it is required that production takes place continuously without interruption.

Flow/Continuous Process Industries

The characteristics of these industries are: (a) they have less processing steps; (b) product routings are same; (c) material flow is divergent and (d) the final product is of less value. Examples of flow process industries are bulk chemical processing industry, glass manufacturing industry, steel moulding industry and paper industry. Some production characteristics of these industries have been discussed by various researchers, which are presented in this section.

Paper and Pulp Industry

Production planning and scheduling in the paper and pulp mill is explained in this section. Santos and Almada-Lobo (2012) described various processes present in the industry manufacturing paper required for cardboard making, from the produced pulp. Initially, wood chips are cooked in a digester which produces two by-products, namely pulp and black liquor. The pulp is mixed with recycled fibres, and is processed in a paper machine. Since various types of paper are produced, significant sequence-dependent setups occur during paper changeover, and therefore lot sizing and sequencing of lots have to be made simultaneously while planning the production of paper. This is because when a switch-over occurs to change the type of paper produced, the new setup leads to loss in production process in terms of time, thereby producing low quality paper at the end of the previous setup. This wasted paper is again fed into the recycled pulp mill. Also, in order to minimize the waste, the sequence is kept in track. In this industry, continuous production of pulp takes place because when chemical pulping is done, wood chips are cooked along with reagents in an aqueous solution under high temperature in a continuous digester. There are complexities in the chemical reaction inside the digester because of different types of wood used, changing the wood feed and speed of rotation of the digester. Due to these reasons, this process should take place continuously without any break in production, even if carried across time periods, because it regulates the pulping quality. Parallel to paper production, forming is another process which takes place continuously. There is a weak black liquor coming out of the digester which goes into an intermediate tank before being concentrated in an evaporator of a given capacity. After getting concentrated, the black liquor goes through a buffer, and then into a capacitated recovery boiler, where it is burnt providing high pressure steam. This steam which is a product obtained after pulping is a by-product which is used for the drying process of paper. It is also sold outside as electrical energy by feeding it continuously into turbines. Hence, the process of production of steam is also done continuously from black liquor without any idle time left in between the steam production process. Therefore, it is understood that both sequence-dependent setups and continuous production without any idle time between production periods are carried out in the paper industry. Poltroniere et al. (2008) explained that sequence-dependent setups occur in paper industry because the customer orders which arrive to the industry differ in width, kind of paper, quantity and due date. Also, the large rolls of paper (jumbos) are produced by different machines, each having a different production rate and different specific timings to change from one type of paper to another. While changing the setup to produce from one type of paper to another, wastage of raw material and machine time causes sequence-dependent setup times to occur. Similarly, Shim et al. (2011) studied the production planning of cardboard from paper waste.

Planning of Beverage Production

The beverage production planning was explained by Camargo et al. (2012). The beverage production is a two-stage production process. Two-stage production process also exists in glass container manufacturing and various other industries. Here, in the first stage, continuous production of a common resource takes place. Continuous production of this common resource goes as an input into the next stage which has a set of parallel machines in which discrete manufacturing takes place, and all the parallel machines process products of the same common resource. In beverage production planning, the first stage consists of preparation of syrup in tanks, and in the second stage, the prepared syrup is distributed parallelly to the bottling machines. The preparation of syrup in tanks has to take place continuously without break, and when produced in batches, the setup should take place each time a new batch is produced, but again the production is done continuously before the next batch is initiated. The production in the first stage (syrup) acts as a setup for the second stage (bottling stage). Therefore, the bottling is synchronized with the syrup production in the first stage. Therefore, the beverage industry is an industry where both discrete manufacturing and continuous manufacturing take place though it is a process industry; and continuous production is carried out in this process industry within and across time periods. Sequence-dependent setups can also occur in beverage industry. Soft drinks are produced in tanks of varying capacities, and when the tank is emptied for a new lot to be produced (even if the flavour is not changed) a significant amount of cleaning time occurs. Sequence-dependent setups also occur when different flavours are produced. In-between two different flavours the cleaning time is more (Ferreira et al. 2012).

Cement Manufacturing Process

In cement manufacturing, continuous production takes place when all the raw materials needed for cement production process are passed through the rotary kiln. Some of the raw materials are blasted limestone, crushed limestone, raw meal, gypsum, red alluvium and clinker (Ogbeide and Omorodion 2016). These are both calcareous and argillaceous raw materials which are crushed and grounded. These grounded particles are stored in a hopper for screening. These materials are mixed in proper proportions and burnt in the rotary kiln by regulating the temperature across the kiln. The burning process which takes place in the kiln is inclined and rotates slowly. The material is passed in the upper end and it moves towards the lower end. Various temperatures are maintained across the kiln. The upper part is maintained around 400°C, for the water or moisture to evaporate. This zone is called drying zone. The central part is maintained around 10000°C where decomposition of lime stone takes place. The lower part is kept around 1500–1700°C where some chemical reactions take place to produce clinkers. Grinding of the clinkers results in cement which are then stored and packed. Since there are different temperatures maintained across same kiln for the process to be carried over, any break in the process of burning will affect the properties of the cement. The cement manufacturing process is shown in Fig. 2.2.

Food Processing Industry

In the food processing industry, the major issue which prevents carrying the setup state from one period to the next is the perishability of most of the products (e.g. milk and yoghurt). Also, idle time cannot be present in between the production periods, and the products have to be dispatched continuously in order to maintain freshness standard (Amorim et al. 2012).

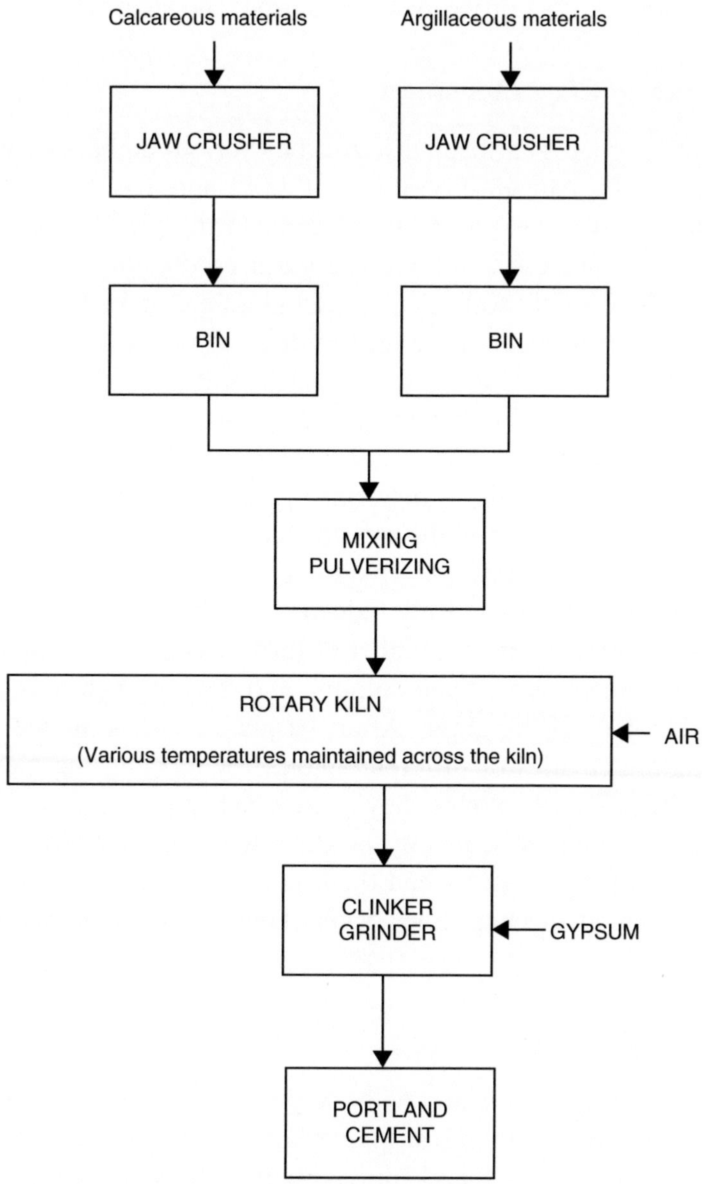

Figure 2.2: Cement manufacturing process

Continuous Casting of Steel Plates

The problem of lot sizing and scheduling in a steel industry is very challenging because of the large number of products produced. Mori and Mahalec discussed about the scheduling of the continuous castings in steel making. The steel plate manufacturing companies have a variety of products to be manufactured with less volume. The manufacturer makes car body parts, housing enclosures for machinery, etc. The large variety makes it difficult to schedule production. The process flow in steel production is as follows: the basic unit of steel production is called as charge and it contains steel of the same grade; these charges are then input to a continuous casting machine where the casting is done from one charge to another charge thereby transforming the molten steel into a solid; once the solid form of steel is produced continuously, they are cut into slabs, and finally the slabs are in turn cut into plates of some specific thickness. Hence, the casting process of steel has to take place continuously without any break to maintain proper final properties of the produced steel slabs.

Planning of Production of Glass Containers

The glass manufacturing process has been discussed by various authors such as Fransoo (1993), Almada-Lobo et al. (2007), Almada-Lobo et al. (2008) and Toledo et al. (2016). Almada-Lobo et al. (2007) studied the glass container manufacturing process, and stated it as a semi-continuous manufacturing process involving the glass production being the continuous part (raw materials such as lime stone, soda ash, broken glass are melted in a furnace at high temperature) and the container production being the discrete part (made in forming machines). Fransoo (1993) describes the processing of glass as a continuous process. Here the raw materials are fed into the furnace. No changeover occurs inside the furnace, the plant operates round the clock (365 days/year), and no break in between the production occurs. If there is a break and the production run ends, a new run has to be set up, the adjustment of the inspection equipment has to be done and the moulds have to be exchanged in the glass forming machine. Almada-Lobo et al. (2008) and Toledo et al. (2016) also addressed the production planning in a glass container manufacturing industry. The industry considered in the study has several facilities, and each facility has a number of furnaces of various capacities. The industry manufactures glass containers of different colours, and the raw materials input to the furnace determines the future colour of the glass. After the glass paste is melted in the furnace, it is passed through moulding cavities of different shapes for forming the glass container. Since there are glass pastes of different colours, they cannot be mixed together. The furnace can only melt one colour at a time and subsequently produce the glass container of that colour. The furnace is operated continuously, and when there is a change over in the colour of the glass paste, large sequence-dependent setup times occur. The glass container manufacturing process is shown in Fig. 2.3.

Process of Sugar Manufacturing

The process of manufacturing sugar starts with the processing of cane. Once the beets arrive at the refinery, they are washed and cut into strips. Following this, they are put into diffusion cells with water at about 175 °F (79.4 °C), and sprayed with hot water counter-currently to remove the sucrose. This process takes place continuously without interruption until all sucrose is removed. After the initial processing of cane, a large part of the juice is extracted via a process

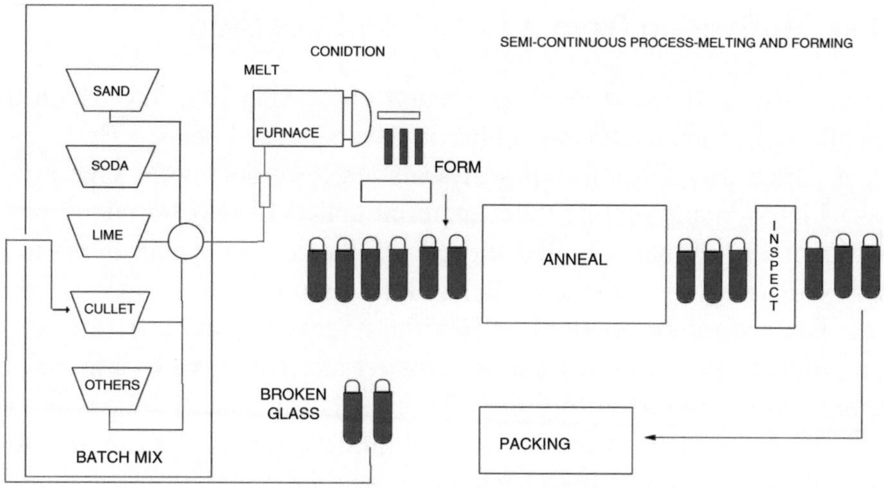

Figure 2.3: Glass container manufacturing process: A schematic diagram

known as pressing. Once the sugarcane juice is extracted, a process known as defecation or clarification takes place to remove the soluble and insoluble impurities from the juice. Once the impurities are extracted, water is removed from the sugarcane juice by a process known as vacuum evaporation. In this process, four vacuum-boiling cells are arranged in series, and each succeeding vacuum-boiling cell has a higher vacuum (which makes the sugarcane juice to boil at a lower temperature). Two-thirds of water from the juice is removed through vacuum-evaporation. This process also takes place continuously without any break removing 35% of water, and the process does not interrupt because the vapour removed in one vacuum-boiling cell is used to boil the incoming syrup (this multiple-effect evaporation continues till the last cell). After this process, the syrup is sent to a machine which runs at 200 RPM for 15 h for the process of crystallization. This process also takes places continuously without any break and the sugar syrup is evaporated until saturated sugar is obtained. The sugar manufacturing process is shown in Fig. 2.4.

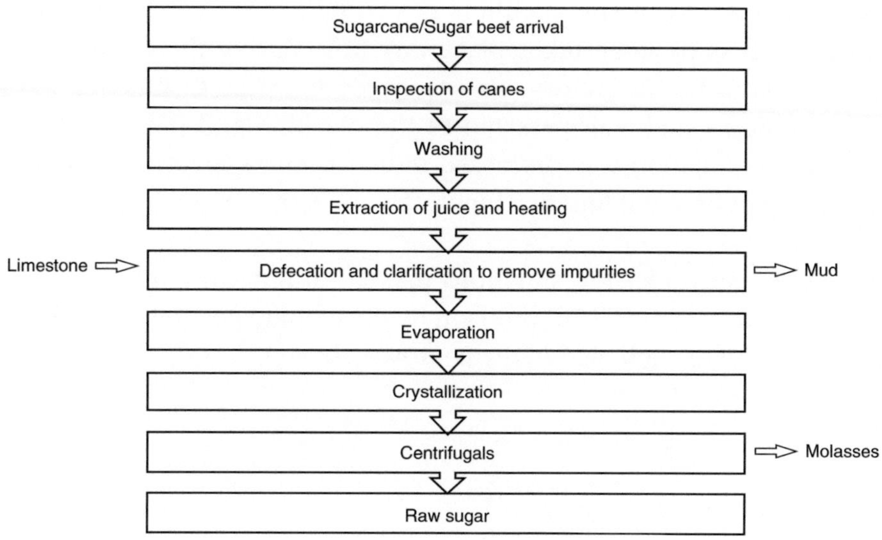

Figure 2.4: Sugar manufacturing process

2.2 Further Motivation from a Real-Life Case Study

A real-life case study has been studied in a batch processing industry which uses injection moulding machines for manufacturing rubber products, called seals. There are many seals produced out of which two different types of seals are discussed here. One product is called the brake seal which is made up of rubber material called EPDM (ethylene propylene diene monomer) and another product is called the oil resistance seal which is made up of rubber material called NBR (nitrile butadiene rubber). Both the products are manufactured using the same machine. The shop floor operates 24 h without any gap, and after the production completion of one product, the setup of the next product starts if there is time left at the end of the day. The process of manufacturing the rubber seals starts with the passing of the rubber material into the hopper which is the inlet of the moulding machine. From the hopper the rubber passes through the screw and barrel which pushes the material into a temporary storage area maintained at 40 °C, known as cold runner block. From the cold runner block the rubber material is pushed into the mould which operates at a temperature higher than the cold runner block. Brake seals are produced using the machine, and after the production completion of the brake seals the operator has to set up the machine to produce the oil seals. The process of setup of the machine takes 5 h. The EPDM rubber which is lying in the hopper, screw and barrel, cold runner block and mould have to be cleaned before starting the production of the oil seals. To clean these parts, the operators inject a changeover white compound to remove all the black sediment particles of rubber present in various parts 2 of the machine. This white compound is passed until the white compound comes out of the machine. Initially the white compound pushes the black sediments of the EPDM rubber due to which the white and black particles get mixed and come out of the machine. In due course, only the white compound comes out of the machine. Next the operators introduce the second material NBR to clean the entire white compound until the black compound comes out of the machine. They do the cleaning process with so much care because the two rubber compounds (EPDM and NBR) should not mix up and if they mix up, the properties of the final product change, which may lead to failure in the braking system (in the case of brake seals) and fuel injection system (in the case of oil seals). This entire process of changeover from black to white to black compound takes about 3 h. Then the mould is heated up to 180 °C before starting to produce the oil seal. The process of setup takes place any time during the day (even when time is left at the end of a day), i.e., operators can set up the machine to produce the next product starting 11 p.m. on day 1 and end the setup of the machine at 3 p.m. on day 2. This continuous manufacturing situation was discussed with the manager of a shop floor that makes rubber seals using injection moulding machines. The injection moulding machine is presented in Fig. 2.5.

In order to address such real-life situations present in process industries, the work in this book is carried out to address the CLSP problem by developing generalized mathematical models and heuristics which are explained in the next two chapters.

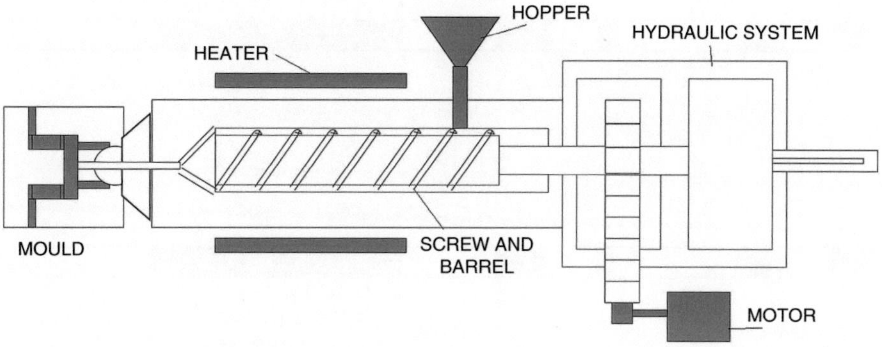

Figure 2.5: Injection moulding machine: A schematic diagram

2.3 Scope of the Book in the Context of Process Industries

Medium term or tactical planning decisions in a production shop floor play an important role in today's competitive and global marketplace. Amongst the various medium term planning decisions, such as sequencing the production runs, lot sizing, overtime assignment to the workers, deciding the workforce level and material requirement planning decisions, lot sizing is one of the most important production planning decisions because it decides the performance of an industry based on satisfying the customer demands, thereby reducing the cost for the industry. Hence, the focus of the present work is lot sizing. The production planning problem or the lot sizing problem decides the quantity of products to be produced, or the number of production orders to be met in a given time period, in order to satisfy the customer's demand at minimum cost. As discussed in the previous sections, lot sizing decisions have been carried out extensively in discrete manufacturing industries as well as continuous manufacturing industries. Studies have also shown various types of lot sizing models proposed in literature (i.e. ELSP, DLSP, PLSP, CSLP and CLSP). However, most production planning decisions have been made using CLSP because of the consideration of time buckets of large sizes which allows the setup and the consecutive production of several products in a period. The CLSP which was initially modelled using setup times and setup costs was later extended to include the carrying of the setup state across periods (thereby reducing the setup cost). This phenomenon is termed as production carryover in this book. Several researchers proposed mathematical models and heuristics to solve this problem (the review of literature on CLSP with production carryover across periods is provided in Chap. 1). The CLSP with production carryover was later extended to include setup splitting between two periods by Mohan et al. (2012) who developed a mathematical model for the problem. However, the mathematical model developed by Mohan et al. (2012) was incorrect leading to infeasible solutions. This was observed by Ramya et al. (2016). They proposed two mathematical models for the CLSP with production carryover and setup splitting. Further, setup crossover across periods along with production carryover across periods was considered by Sung and Maravelias (2008) and Belo-Filho et al. (2013). These authors developed mathematical models considering setup crossover across periods, in order to address production situations in continuous manufacturing industries such as process industries having long setup times. The differences in the terminology used in this book versus the work by Belo-Filho et al. (2013) are illustrated in Fig. 2.6. Figure 2.6a illustrates the setup crossover (period-overlapping setup), as defined by Belo-Filho et al. (2013), is the opportunity to start

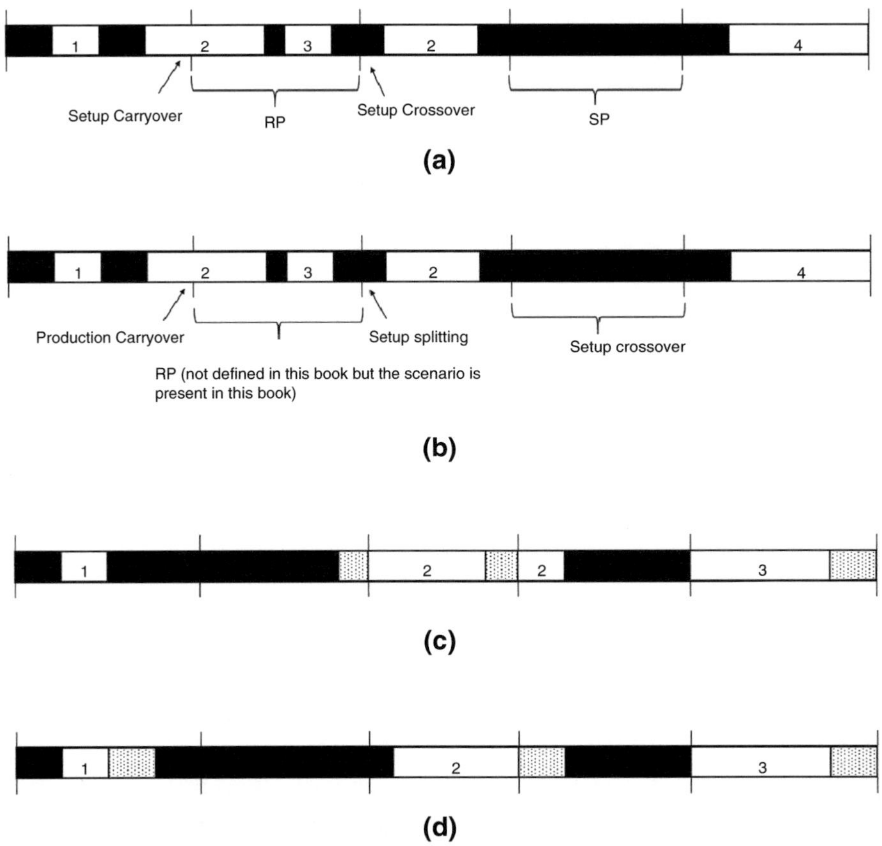

Figure 2.6: Solution generated by model of Belo-Filho et al. (2013) illustrating the RP and SP scenario vs solution generated by our proposed mathematical models. See Sect. 3.5 in Chap. 3 for further discussion. (**a**) Gantt chart illustrating a possible solution of Belo-Filho et al. (2013) along with the terminology used in their paper. (**b**) Gantt chart illustrating a solution of proposed mathematical models along with the terminology used in this book. (**c**) Gantt chart illustrating a possible solution of Belo-Filho et al. (2013). (**d**) Gantt chart illustrating a solution of the proposed mathematical models in this book

a setup operation in one period and continue it to the following one, that is, the incomplete setup operation crosses over time period boundaries. In case of long setup times (in relation to the size of the period, may be even greater than one period length), the setup operation may be performed in more than one full period. This scenario where a setup time is longer than a complete period is when the setup starts in a period and finishes in one of the following periods. Therefore, an entire period (or more) may be dedicated to an in-progress setup operation. This setup in progress is termed as SP scenario by Belo-Filho et al. (2013). The same phenomenon where a setup gets initiated in one period and can extend across one or more than one period is defined as setup crossover in this book (see Fig. 2.6b). In this book we define the phenomenon of a setup starting in one period and completing in the next period as setup-splitting (similar to Ramya et al. (2016)). Since this book addresses scenarios in process industries which have predominantly long setup times the term setup crossover is used (similar to SP scenario of Belo-Filho et al. (2013)) to address long setups. The phenomenon setup crossover also includes the concept of setup splitting between two periods. When the setup is less than the capacity of a period it is not an SP scenario similar to that of Belo-Filho et al. (2013), rather it is a setup splitting scenario. However, the term setup crossover is used throughout this book to describe any setup which spans across periods.

In the paper by Belo-Filho et al. (2013), the authors defined another scenario (see Fig. 2.6a) where the setup state of an item is to be active twice in the same period, which is forbidden or cost-prohibitive in the CLSP-SC problems; that is, there is a return to the initial product setup state (return product or RP scenario). The same condition applies to the mathematical models proposed in this book. However, a terminology to define the RP scenario is not present in this book (see Fig. 2.6a). Carrying the setup state across periods is termed as setup carryover by Belo-Filho et al. (2013) (Fig. 2.6a) and it is termed as production carryover in this book (Fig. 2.6b).

Figure 2.6d shows that in case of process industries there should not be any break in-between the setup and the consecutive production of the product. Also in process industries the production should take place continuously without any interruption across periods. Otherwise, the properties of the final product may change (shape, texture, thickness etc.). Figure 2.6c shows a possible solution by the model of Belo-Filho et al. (2013). The model does not address production immediately after setup and uninterrupted production carryover across periods which are conditions to be satisfied in process industries (also see Chap. 3, Sect. 3.5 for further details).

Belo-Filho et al. (2013) stated that if the planning horizon of the problem is treated as continuous (for instance, 24/7 industrial environments), small-bucket and large-bucket formulations that do not assume setup crossover do not take into account all possible solutions of the feasibility domain. Furthermore, without the setup crossover feature, the decision maker is not totally free to choose the period size, which, in this case, would have to be at least the size of the longest setup time. Further, their paper states that the continuity of scheduling decisions across periods, for production and setup operations, respectively, can be incorporated by assuming the production carryover and crossover assumptions. Such assumptions are appreciated, for instance, in process industries with considerable setup times. Setups in process industries usually deal with extensive cleansing-up operations. Testing operations should be performed to guarantee that no contamination affects the following processes. Therefore, setup times consume a significant part of the period's length, augmenting the importance of making a flexible assignment and timing of the production and setup operations. Production carryover and crossover are applicable in chemical and beverage industries (Sung and Maravelias 2008). However, the models developed by Sung and Maravelias (2008) and Belo-Filho et al. (2013) did not address real-life situations present in most process industries, such as production starting immediately after setup and uninterrupted production carryover across periods, along with the presence of long setup times. Hence, the models proposed in this book are more generalized that they can handle setups crossing one or more periods depending on the industry. They can address the case of production immediately after setup and uninterrupted production across periods. The models are explained in detail in the chapters to follow.

The work by Mustafa and Cheng (2017) explains the presence of long setups with an industrial case study where a food manufacturing company produces three flavours of snacks in a single production line. When there is a change over between products (snacks), the setup time for the next product is long. For instance, the first machine (oil spray drum) takes 45 min to be cleaned and there are some manual wipes and reassembly which needs to be done. The second machine (flavour adding machine) takes 150 min to clean as cleaning different flavours become difficult. The cleaning is done manually and even after cleaning the parts are assembled back to the machine for the next production run. The cleaning time also depends on the worker and their experience. There is also a storage conveyor which takes 240 min approximately to clean. It is also observed that 16.50% of downtime is attributed to changeover and 23.23% of the downtime is on account of cleaning the machines and conveyors. Both contribute 39.73% to the total downtime. Apart from the changeover and cleaning downtimes, the paper explains that there is also downtime which ranges to several days and setup the machine for the next production run. This is one more real life case to show that a setup can take more than a time bucket (shift/days). Further, Sect. 2.1 provides a detailed explanation of these real-life characteristics of continuous manufacturing industries such as process industries. It can also be seen that most process industries have the presence of sequence dependent setups combined with the presence of long setup times.

This book is primarily motivated by the industrial practices that are widely reported in literature. In the case of a discrete manufacturing industry, setup splitting is possible especially when the production of a product is completed earlier in a day; e.g., in the last shift of a day and there is time left in the current shift to set up the next product to be produced. This is done by initiating the setup of the next product in the current day's last shift and continuing the setup 1 of the same product in the next time period's (next day's) first shift. The setup splitting can be performed across two periods (two different days) when done by operators having the same skill set. In the case of a discrete manufacturing industry, mostly there is no necessity that the production should be continuous without interruption. On the other hand in continuous manufacturing industries such as most process industries, such conditions will necessarily hold.

The problem statement in this book refers to process industry situations where the time of setup runs in the order of hours/shifts or days. Referring to the benchmark papers that are related to our study (e.g., Sung and Maravelias (2008) and Bclo-Filho et al. (2013)) we find that the setup time may extend beyond one time bucket and the setup time is comparable or perhaps more than the production run time (i.e. the time for setup may extend beyond a shift or a day). This is especially true in case of process industries such as chemical, sugar, cement, metallurgical industries. The situation where the setup time is possibly more than the production run time holds good in the case of high value and process intensive industries such as pharmaceutical industries where the production quantity per unit time fetches high revenue in comparison with the cost incurred with respect to long setup duration (due to high technology intensive setups associated with process controls and monitoring while setting up the process for manufacturing, for the production of end products associated with high cost raw materials and high revenue/unit sales end products). The paper by Berndt (2002) explains how it takes around 12 years to bring a new medicine to market. Apart from the basic R&D, a large part of it goes to the clinical costs of development. Hence the launch and production of the medicine starts only after intensive research on development of the new drugs. Paul et al. (2010) explained new drugs development and the number of years and cost involved before launching the new medicine in the market.

However, the models proposed in this book are not constrained by the consideration of long setup products. The models are flexible enough to handle the process industries with small bucket setups and long bucket production runs or the scenario with large bucket setups and small production runs or a mixture of both. In other words, the proposed mathematical models and heuristic approaches are flexible enough to handle or address situations in the conventional process industries such as cement and sugar industries (associated with small bucket setups and long bucket production runs), large bucket setups and small bucket production runs (associated with highly technological intensive big bucket setups and small bucket production runs such as those in highly specialized pharmaceutical processes), or a mixture of scenarios in a single process industry. It is to be noted that in all these scenarios we have real life restrictions that once a process starts, there is no interruption with the production run length, and the production has to start immediately after the completion of setup. In this book we address such a variety or a mix of process-industry scenarios and the real-life restrictions in terms of continuous production and production commencement immediately after setup. This book is also motivated by the literature on CLSP based on the nature of continuous manufacturing industries such as chemical manufacturing, cement manufacturing, sugar industries, pharmaceuticals, hot rolling process, heat treatment, casting and injection moulding, and a real-life case study in a batch processing industry. Referring to the benchmark literature (e.g. Sung and Maravelias (2008) and Belo-Filho et al. (2013)), we find that no existing work has attempted such a mix of industrial scenarios and associated real-life constraints such as the continuous production of a product with no interruption and production commencement immediately after its setup completion.

2.4 Summary

The general production planning situations in discrete manufacturing and continuous manufacturing industries is discussed in this chapter. With the ideas obtained from the CLSP literature, in order to address the production situations present in almost all process industries, the motivation and scope of the current study has been presented. Most of the studies in CLSP have developed mathematical models and heuristics for problems considering a single machine or multiple machines, considering a single level or multiple levels of a product. Authors have also developed mathematical models and solution procedures to address the CLSP for various industries which required production planning, thereby minimizing the setup and holding costs, but most authors failed to focus on the effective utilization of capacity of a period. Mohan et al. (2012) and Ramya et al. (2016) extended the CLSP with production carryover across periods by including setup-splitting between two periods in order to use the capacity efficiently, and addressed the problem by developing a mathematical model. Setup splitting is a phenomenon where the leftover capacity at the end of a period t can be used to set up a product which gets completed in period $t + 1$. Few researchers also addressed the presence of long setup times present mostly in process industries. Sung and Maravelias (2008) and Belo-Filho et al. (2013) addressed the CLSP with production carryover across periods, along with the presence of long setup times thereby developing mathematical models. Belo-Filho et al. (2013) developed two mathematical models for this CLSP problem considering backlogging and setup crossover. Mathematical models and heuristics have been proposed in the literature addressing CLSP with sequence-dependent setups, with the presence of production carryover (known as CLSD in most of the papers), and without setup crossover, some of which are the Lagrangian decomposition methods, meta-heuristics, hybrid Lagrangian-simulated annealing based heuristic algorithm and the fix-and-optimize approach. The presence of long setup times in process industries and real-life situations such as the production starting immediately after the setup of a product and uninterrupted production carryover has not been addressed in earlier literature. This scenario which is the focus of our study is motivated by continuous manufacturing in real-life batch process industries. Considerations of sequence-dependent setup times and the associated costs in real-life process industries have also been addressed in this book.

CHAPTER 3

Capacitated Lot Sizing Problem with Production Carryover and Setup Crossover Across Periods (CLSP:PCSC): Mathematical Model 1 (MM1) and a Heuristic for Process Industries

3.1 Introduction and Problem Definition

The capacitated lot sizing problem (CLSP) is a lot sizing model in which the production of multiple products is allowed within a time period on a single machine, with a condition that the entire demand for a product within that period should be met from the production in that period and/or the inventory carried from the previous periods, without any backorders or lost sales. Finding a minimum cost production plan that satisfies all the demand requirements without exceeding the capacity limits of a period is the main objective of the CLSP.

Research has been carried out in the area of CLSP and is extended to include production carryover across periods. Haase and Drexl (1994), Gopalakrishnan et al. (1995), Sox and Gao (1999), Gopalakrishnan (2000) and Suerie and Stadtler (2003) are some of those who addressed the CLSP with production carryover by proposing mathematical models. Further, researchers also proposed several approaches to solve the CLSP with production carryover by developing algorithms and heuristics. Some of them are due to Gopalakrishnan et al. (2001), Suerie and Stadtler (2003), Karimi et al. (2006), Nascimento and Toledo (2008), Caserta et al. (2009), Sahling et al. (2009), Caserta et al. (2010), Goren et al. (2012), Wu et al. (2013) and Caserta and Voß (2013). All the aforementioned attempts basically assume that while the production of a product can be carried over across periods, they have assumed that the setup cannot be performed across time periods; i.e., setup, once started, needs to be completed within the same time period, and the spill-over to the next time period is not allowed.

When the capacity of a machine has to be utilized efficiently, the idle time present in a period should also be utilized judiciously. Therefore, there are three ways by which a machine can be set up for producing a product when $max_i\{ST_i\} \leq C_t$, $\forall t$. They are: (i) a machine is completely set up for product i anywhere in period t, and the production starts in period t itself and the production may be continued to the period(s) thereafter (this aspect is referred to as production carryover in this book); (ii) when there is enough capacity left at the end of a period, it can be utilized in making a complete setup for a product, followed by its production in time period $t + 1$ and (iii) the setup started in period t can be split between the periods t and $t + 1$, followed by its production in time period $t + 1$ (this aspect is referred to as setup splitting in this work). These are the three ways of setting up a machine for production. The

R. Ramya et al., *Capacitated Lot Sizing Problems in Process Industries*, https://doi.org/10.1007/978-3-030-01222-9_3

figures corresponding to the three ways are shown in Fig. 3.1. The X-axis in the Gantt chart denotes the time periods represented as 1, 2, 3 and so on. The shaded region indicates the setup and the blank region indicates the production. This CLSP with production carryover and setup splitting (with no backorders or lost sales) was addressed by Mohan et al. (2012) and Ramya et al. (2016) with the help of mathematical models.

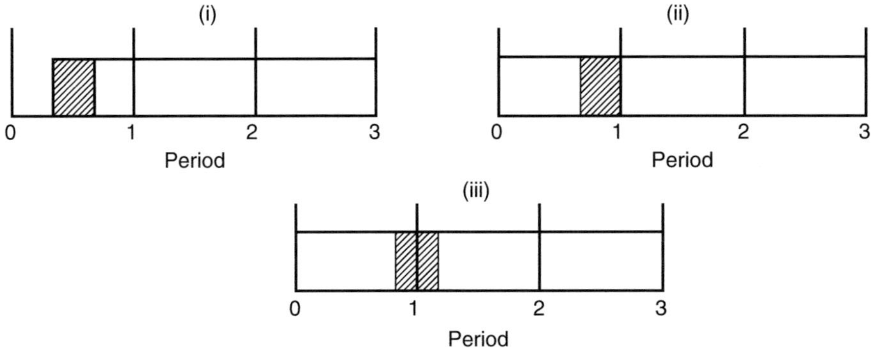

Figure 3.1: Three ways of a machine being set up in a period for production when $max_i\{ST_i\} \leq C_t, \forall t$

Long setup times (possibly spanning across consecutive periods) are prevalent in continuous manufacturing industries such as process industries. In industries where there are long setup times, manufacturers allow the setup to be carried over across periods. Allowing production carryover, setup splitting and setup crossover across periods help the manufacturer to use the capacity effectively. The problem statement in our study refers to process industry situations where the time of uninterrupted setup spans over hours/shifts/days or weeks. Referring to the benchmark papers that are related to our study (e.g. Sung and Maravelias (2008) and Belo-Filho et al. (2013)), we find that the setup time may extend beyond one time bucket and the setup time is comparable or perhaps more than the production run time (i.e. the time for setup may extend beyond a shift or a day). This is especially true in case of process industries such as chemical, sugar, cement and metallurgical industries. The situation where the setup time is possibly more than the production run time holds good in case of high value and process intensive industries such as pharmaceutical industries where the production quantity per unit time fetches high revenue in comparison to the cost incurred with respect to long setup duration (due to high technology intensive setups associated with process controls and monitoring while setting up the process for manufacturing, for the production of end products associated with high cost raw materials and high revenue/ unit sales of end products).

However, the proposed model in this chapter is not constrained by the consideration of long setup products. The model is flexible enough to handle the process industries with small bucket setups and long bucket production runs, or the scenario with large bucket setups and small production runs, or a mixture of both. In other words, the proposed mathematical models and heuristic approach are flexible enough to handle or address situations in the conventional process industries such as cement and sugar industries (associated with small bucket setups and long bucket production runs); large bucket setups and small bucket production runs (associated with highly technological intensive big bucket setups and small bucket production runs such as those in highly specialized pharmaceutical processes); or a mixture of scenarios in a single process industry. Also, depending upon the industry, the definition of a period (measured in terms of shifts or days) may vary. It is to be noted that in all these scenarios, we have real-life restrictions that once a process starts there is no interruption with the production run, and the production has to start immediately after the completion of the uninterrupted setup. In this book we address such a variety of process-industry scenarios and the restriction in terms of continuous production and production commencement immediately after setup. This book is motivated by the literature on CLSP based on the nature of continuous manufacturing industries such as chemical manufacturing, cement manufacturing, sugar industries, pharmaceuticals, hot rolling process, heat treatment, casting and injection moulding, and a real-life case study in a batch processing industry. Referring to the benchmark literature (e.g. Sung and Maravelias (2008) and Belo-Filho et al. (2013)), we find that no existing work has attempted such a mix of industrial scenarios and associated real-life constraints such as continuous production with no interruption and production commencement immediately after setup completion. Therefore, the proposed mathematical models in this book are more generalized and comprehensive in nature.

There can be three ways by which a machine is set up for producing a product. They are: (i) a machine is completely setup for product i anywhere in period t, and the production starts in period t itself and the production may be continued to the period(s) thereafter (production carryover); (ii) when there is enough capacity left at the end of a period, it can be utilized in making a complete setup for a product, followed by its production in time period $t + 1$. An end-of-period setup of product i can occur when an amount of capacity is left at the end of period t to setup the product, or it can also refer to the setup which takes a value equal to the capacity of the period in which its setup is initiated and (iii) the setup started in period t can be carried over to period t', followed by its production in time period t' or $t'' > t'$ depending upon where the setup of product i ends (this aspect is referred to as setup crossover in this book). The phenomenon setup crossover also includes the concept of setup splitting between periods t and $t + 1$, when the setup time of product i is less that the capacity of period t in which its setup is initiated, and less than the capacity of period $t + 1$ in which its setup is completed. The figures corresponding to the three ways of setup are shown in Fig. 3.2.

The presence of idle time of a production resource during the setup or production of a product or the production after its setup completion of a product is not acceptable in the case of almost all process/chemical industries. Therefore, solutions generated by lot sizing models cannot have the presence of idle time of the production resource between production periods as they are not acceptable solutions for process industries. In these industries, the production should start immediately after the setup, and no break during the course of production of a product is allowed. Such situations are present in industries such as hot rolling process and heat treatment, chemical manufacture, cement manufacture, sugar industry and pharmaceuticals. In continuous manufacturing such as industries involving heat treatment and hot rolling process, and in almost all process industries, the idle time cannot be present during the course of production. Some examples pertaining to these rigid conditions are now briefly presented.

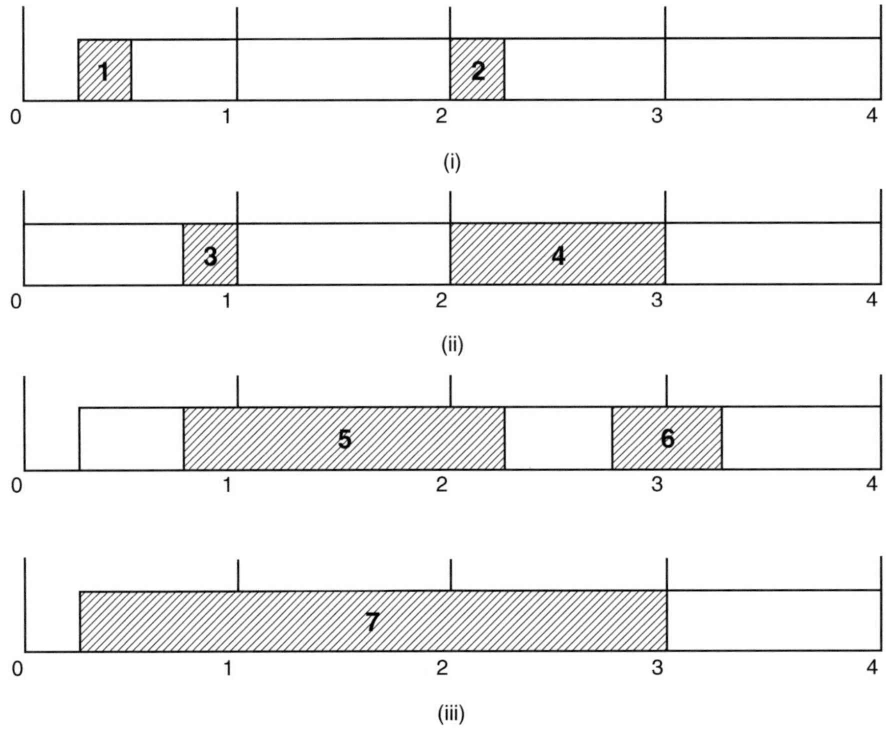

Figure 3.2: Three ways of a machine being set up in a period for production in process industries

Remark: In Fig. 3.2, the X-axis in the Gantt charts denotes the time periods represented as 1, 2, 3 and so on. The shaded region indicates the setup and the blank region indicates the production. Figure (i) indicates a complete setup for product 1 in period 1, with its production starting in period 1 and continued to period 2; another complete setup for product 2 is initiated in period 3, with its production starting in period 3 and continued to period 4 (this setup situation is indicated by $\delta^1_{i,t}$ in MM1:CLSP-PCSC). Figure (ii) indicates an end-of-period setup for product 3 in period 1, with its production starting in period 2; and another end-of-period setup for product 4 in period 3, with its production starting in period 4 (this setup situation is indicated by $\delta^2_{i,t}$ in MM1:CLSP-PCSC). The first Gantt chart in (iii) indicates a setup crossover of product 5, starting in period 1 and carried over to period 3, followed by its production in time period 3; and another setup crossover of product 6, starting in period 3 and ending in period 4, followed by its production in period 4 (this setup situation is indicated by $\delta^3_{i,t,t'}$ in MM1:CLSP-PCSC). The second Gantt chart in (iii) indicates a setup crossover of product 7, starting in period 1 and carried over to period 3, ending exactly at the end of period 3, followed by its production in time period 4 (this setup situation is indicated by $\delta^4_{i,t,t'}$ in MM1:CLSP-PCSC).
Note: Description of all the decision variables is provided in Sect. 3.3.2.

First the hot rolling process is considered. When the continuously cast steel slabs are processed in the hot rolling line, they have to be heated continuously in the reheating furnace at a high temperature and any change in the temperature conditions due to shutting down (or idle time of furnace) triggers the oxidation of the metal. The slabs also have to be passed through the roughing mill and the finishing mill, continuously without interruption, to maintain the dimensions of the final rolled steel with correct metallurgical and chemical properties. Hence the entire production process should take place continuously without any break (i.e. without any idle time of the production resource during the course of production across time periods), and if there is a break in the production, the properties of the final product will get affected. The production resource can remain idle after the completion of production of a product and before the commencement of setup of the next product. It should be noted that if there is any discontinuity or break in the production across periods, a new setup is to be done for the same product for any subsequent production, in order to maintain the composition and chemical properties of the product.

Similar production conditions can be found in process industries such as sugar and cement industries. In cement industry, when the raw materials (clay and limestone) are burnt in the rotary kiln under high temperature present across the kiln; in order to obtain the proper chemical properties of the final product, namely cement, the process should be carried out continuously without any break. In sugar industry, the sugarcane juice is passed through evaporators having various pressure settings to remove all the moisture from the sugar syrup and this process also should be carried out continuously without any break.

The next important condition is that the production commencement should immediately follow the setup completion. Ventura et al. (2002) discussed about the preheating time required by the aluminium ingots in steel production. The ingots are preheated up to the required temperature in gas soaking pits before it gets hot-rolled by a blooming mill. Here, the preheating time of an ingot is treated as the release time at which the ingot is available for rolling. Therefore in this scenario, the production commencement (hot rolling process) is done immediately after the setup completion (heating of the aluminium ingot). In die casting industry, heating the furnace of the machine up to the right temperature is the process of setup for casting. Once the furnace is heated, the metal (such as aluminium or zinc) is melted and the molten metal is passed through the mold/die under high pressure to produce the castings, in order to obtain the proper shape of the final product. If there is a break in between the heating of the furnace for melting the metal and the production of the castings, the metal starts getting solidified even before it passes through the casting mold/die. Similarly, in the injection moulding machine, heating the injection unit of the injection moulding machine up to the right temperature is the process of setup. The raw material (plastic raw material granules) is heated in the injection unit until the molten state, and the injection of the molten material into the mold is done immediately so that the production process starts thereafter. Similarly, in pharmaceutical industries, production starting immediately after the completion of setup and uninterrupted production should be carried out because maintaining the proper chemical composition and properties of the those products are mandatory.

Therefore, there is a need to develop mathematical models that can address real-life situations arising across such process industries, and hence to solve the CLSP in those industries, i.e., production to commence immediately after setup; and no break or discontinuity during the course of production. The presence of long setup times is also considered while developing the models. The setup time of a product is allowed to take a value greater than or equal to the capacity of a period and the setup is carried over from one period to another when the product has a long setup time. Belo-Filho et al. (2013) had already worked on the CLSP with setup crossover and production carryover across periods, allowing the presence of backorders. These authors considered the presence of long setup times. However, they did not address real-life situations prevailing in most process industries such as production commencement immediately after the completion of setup and no break or discontinuity during the course of production (discussed in detail later).

Therefore, the main contribution of this chapter is to address this gap, by proposing a mathematical model and a heuristic. A generalized mathematical model Mathematical Model 1 (MM1:CLSP-PCSC) is developed for the capacitated lot sizing problem with production carryover and setup crossover across periods (CLSP-PCSC) which can address real-life situations present in almost all continuous manufacturing industries such as process industries. The mathematical model can handle situations where the setup costs and holding costs are time independent, as well as situations where the costs are time dependent. The time-dependent structure is considered in view of the well-known Wagner–Whitin algorithm for the dynamic lot sizing problem (Wagner and Whitin 1958) which addresses the scenario where the cost of setup varies from one time period to another. A heuristic based on MM1:CLSP-PCSC is also proposed, in view of the basic CLSP being NP-hard (Bitran and Yanasse 1982).

This chapter is organized as follows: The basic assumptions of the proposed mathematical model MM1:CLSP-PCSC are presented in Sect. 3.2. Section 3.3 presents the parameters and decision variables for MM1:CLSP-PCSC (see the corresponding Sects. 3.3.1 and 3.3.2, respectively). The generalized version of mathematical model MM1:CLSP-PCSC considering the scenario where the setup cost of a product is calculated with respect to the period of initiation of setup of the product is presented in Sect. 3.3.3. Method of tracking setups in MM1:CLSP-PCSC is presented in Sect. 3.3.4. In Sect. 3.4, special cases of MM1:CLSP-PCSC are presented with appropriate changes made to the objective function and add-on constraints. They are: MM1:CLSP-PCSC when the setup cost of a product is calculated with respect to the period of its setup completion (Sect. 3.4.1); MM1:CLSP-PCSC when the setup cost and holding cost of a product are time independent (Sect. 3.4.2). Following this, Sect. 3.5 presents the numerical illustration and discussion for the generalized MM1:CLSP-PCSC and its special cases. Here, some observations from the model developed by Belo-Filho et al. (2013) are explained with an example. A heuristic is proposed with respect to MM1:CLSP-PCSC, and implemented for solving the CLSP-PCSC (see Sect. 3.6). The computational time for generalized MM1:CLSP-PCSC is presented in Sect. 3.7.1. The comparison of the heuristic and the exact method of MM1:CLSP-PCSC is presented in Sect. 3.7.2. The chapter is concluded with a summary of the contributions and is presented in Sect. 3.8.

3.2 Basic Assumptions of the Proposed Mathematical Model (MM1: CLSP-PCSC)

- A single machine is considered in the problem.
- Multiple products can be produced on the single machine and each product is made up of a single level.
- Time unit is discrete and the time horizon considered is finite.
- Each product is associated with a setup cost when set up on the machine, and it consumes time for setup.
- Backorders are allowed but lost sales are not permitted.

- The capacity of the machine during a given period is assumed in time units and it may vary from period to period.
- The capacity of the machine per period is consumed by the setup time and the production time of the products. Idle time on the machine can also be present.
- If excess capacity is left over on the machine in a period after production in period t, it may be used to setup the product to be produced in the next period. If the setup is not over, this setup may be continued to some future period t', where $t' > t$. In this work this aspect is called setup crossover.
- The excess quantity produced of a product can be stored and this incurs a holding cost, except in the last period where all the units in the inventory have to be consumed.
- Production of a product may extend over any number of periods subject to demand and capacity constraints. In this work, this aspect is called production carryover.
- At most one setup of a given product i can be initiated in the given time period t. It means that the carryover of a setup is permitted from one of the previous time periods to the present period, and the initiation of setup of the same product in that given time period is also permitted.

3.3 Mathematical Model (MM1:CLSP-PCSC) for the Capacitated Lot Sizing Problem with Production Carryover and Setup Crossover Across Periods

The mathematical model (MM1:CLSP-PCSC) can handle situations where the setup costs and holding costs are product dependent and time dependent. In the classical Wagner–Whitin algorithm (Wagner and Whitin 1958) for the dynamic lot sizing problem, time-dependent cost structures such as time-dependent setup (order) costs and holding costs were considered. In this section, the parameters (Sect. 3.3.1), decision variables (Sect. 3.3.2) and the generalized Mathematical Model 1 (MM1:CLSP-PCSC) which considers time-dependent cost situations similar to the Wagner–Whitin algorithm (Sect. 3.3.3) are presented. The scenario where the setup cost is calculated with respect to the period of setup initiation is considered while presenting the generalized mathematical model (MM1:CLSP-PCSC) presented in Sect. 3.3.3.

3.3.1 Parameters/Indices

N	number of products
T	number of time periods
t	a given time period
i	product
$SC_{i,t}$	setup cost for product i, when its setup is initiated in period t; this cost is incurred only once as a fixed cost computed with respect to the period of its setup initiation
b_i	backorder cost per period per unit of product i

$h_{i,t}$	holding cost per period per unit of product i in period t
ST_i	setup time for product i
a_i	number of time units required for producing one unit of product i
C_t	capacity of the machine in period t (in time units)
$d_{i,t}$	demand for product i in period t
M	a large value
$\mathcal{E}$	smallest unit of time
$\mathcal{E}_d$	unit of smallest quantity of production

3.3.2 Decision Variables

Variable	Description
$\delta^1_{i,t}$	An indicator (binary) variable that takes value 1 if a complete setup is done for product i in period t with the production starting in period t; 0 otherwise.
$\Delta^1_{i,t,t'}$	An indicator (binary) variable that takes value 1: it corresponds to the production carryover from period t to period t' ($t \leq t' \leq T$), due to the setup of product i started and finished in period t, with no intermittent setup of any other product; 0 otherwise.
$\delta^2_{i,t}$	An indicator (binary) variable that takes value 1 if a setup of product i is started and completed exactly at the end of period t, followed by its production starting in period $t+1$; 0 otherwise.
$\Delta^2_{i,t,t'}$	An indicator (binary) variable that takes value 1: it corresponds to the production carryover from period t' to period $t'+1$ ($t+1 \leq t' \leq T$), due to the end-of-period setup of product i in period t, with no intermittent setup of any other product; 0 otherwise.
$\delta^3_{i,t,t'}$	An indicator (binary) variable that takes value 1 if the setup of product i is commenced in period t and is completed during some period t' but not exactly at the end of period t' ($t+1 \leq t' \leq T$); 0 otherwise.
$\Delta^3_{i,t,t',t''}$	An indicator (binary) variable that takes value 1: it corresponds to the production in period t'' ($t' \leq t'' \leq T$), due to the setup of product i initiated in period t and completed in period t' but not exactly at the end of period t' ($t+1 \leq t' \leq T$) and with no setup of any product during the intermittent periods from period t' to period t''. *Note:* This variable corresponds to (i.e. indicates) the production carryover through periods t' and t'', after the completion of setup in period t' (but not exactly at the end of period t'), with the initiation of setup in period t; 0 otherwise.

Variable	Description
$\delta^4_{i,t,t'}$	An indicator (binary) variable that takes value 1 if the setup of product i is commenced in period t and is completed exactly at the end of period t' $(t+1 \leq t' \leq T-1)$; 0 otherwise.
$\Delta^4_{i,t,t',t''}$	An indicator (binary) variable that takes value 1: it corresponds to the production in period t'' $(t'+1 \leq t'' \leq T)$, due to the setup of product i initiated in period t and completed exactly at the end of period t' $(t+1 \leq t' \leq T-1)$ and with no setup of any product during the intermittent periods from period t' to period t''. *Note:* This variable corresponds to (i.e. indicates) the production carryover through periods t' and t'', after the completion of setup exactly at the end of period t', with the initiation of setup in period t; 0 otherwise.
$I_{i,t}$	Inventory of product i at the end of period t.
$B_{i,t}$	Backorder quantity of product i at the end of period t.
$s^1_{i,t}$	Setup time of product i in period t that takes the value of ST_i, and associated with $\delta^1_{i,t}$.
$s^2_{i,t}$	Setup time of product i in period t that takes the value of ST_i, and associated with $\delta^2_{i,t}$.
$s^3_{i,t,t',t''}$	Setup time of product i in period t'' $(t \leq t'' \leq t')$, when its setup has started in period t and completed in period t', but not completed exactly at the end of period t', and associated with $\delta^3_{i,t,t'}$; *Note:* $\sum_{t''=t}^{t'} s^3_{i,t,t',t''} = ST_i$.
$s^4_{i,t,t',t''}$	Setup time of product i in period t'' $(t \leq t'' \leq t')$, when its setup has started in period t and completed exactly at the end of period t', and associated with $\delta^4_{i,t,t'}$; *Note:* $\sum_{t''=t}^{t'} s^4_{i,t,t',t''} = ST_i$.
$X^1_{i,t,t'}$	Production quantity of product i in period t' (due to its setup started and ended in period t), with $1 \leq t \leq T$ and $t \leq t' \leq T$, and associated with $\Delta^1_{i,t,t'}$.
$X^2_{i,t,t'}$	Production quantity of product i in period t' (due to its setup started in period t and completed exactly at the end of period t), with $1 \leq t \leq T-1$ and $t+1 \leq t' \leq T$, and associated with $\Delta^2_{i,t,t'}$.
$X^3_{i,t,t',t''}$	Production quantity of product i in period t'' (due to its setup started in period t and ended in period t' but not completed at the end of period t'), with $1 \leq t \leq T-1$, $t+1 \leq t' \leq T$ and $t+1 \leq t'' \leq T$, and $t'' \geq t'$, and associated with $\Delta^3_{i,t,t',t''}$.
$X^4_{i,t,t',t''}$	Production quantity of product i in period t'' (due to its setup started in period t and completed at the end of period t'), with $1 \leq t \leq T-2$, $t+1 \leq t' \leq T-1$ and $t+2 \leq t'' \leq T$, and $t'' > t'$, and associated with $\Delta^4_{i,t,t',t''}$.

3.3.3 Mathematical Model 1 (MM1:CLSP-PCSC)

In this section, the mathematical model (MM1:CLSP-PCSC) is presented, with an objective of minimizing the time-dependent setup costs, holding costs and backorder costs of all products across all time periods. The proposed mathematical model assumes that the setup cost of a product is calculated with respect to the period of its setup initiation. This mathematical model helps to address production situations in process industries. They address situations such as presence of long setup times, production starting immediately after the product's setup and uninterrupted production carryover across periods. In this mathematical model, four binary variables are used to track four types of setup, i.e., a complete setup done (started and finished) anywhere in period t (but not exactly at the end of period t), an end-of-period setup, a setup crossing over across a number of periods with the setup initiated anywhere in period t and ending in the middle of some future period t', and a setup crossing over across a number of periods with the setup initiated anywhere in period t and completed exactly at the end of period t'. These binary variables which indicate a product's setup are in turn linked with production carryover indicator variables that help to track the time period in which the product's setup is initiated and completed, and the time period in which the corresponding production is carried out. Production variables corresponding to these production carryover indicator variables are also present which help to determine the production time (here, the production quantity is measured in time units). There are variables which determine the setup time of a product in a period. When the setup of product i crosses over a number of periods, the time taken to setup product i is also split across these periods. There are also constraints to ensure that production starts immediately after setup and uninterrupted production takes place across periods. Through this mathematical model, it is ensured that the demand for all products is satisfied across the entire time horizon with the condition that the production time and the setup time of all the products setup in a period do not exceed the capacity limitations (measured in time units) of that period.

Objective Functional

$$\text{Min} \quad Z = \sum_{i=1}^{N}\sum_{t=1}^{T} SC_{i,t}\,\delta^1_{i,t} + \sum_{i=1}^{N}\sum_{t=1}^{T-1} SC_{i,t}\,\delta^2_{i,t} + \sum_{i=1}^{N}\sum_{t=1}^{T-1}\sum_{t'=t+1}^{T} SC_{i,t}\,\delta^3_{i,t,t'} + \tag{3.1}$$

$$\sum_{i=1}^{N}\sum_{t=1}^{T-2}\sum_{t'=t+1}^{T-1} SC_{i,t}\,\delta^4_{i,t,t'} + \sum_{i=1}^{N}\sum_{t=1}^{T} h_{i,t}\, I_{i,t} + \sum_{i=1}^{N}\sum_{t=1}^{T} b_i\, B_{i,t}$$

Subject to the Following:

/* Constraints (3.2)–(3.7) represent the conditions for setting up a product only once in period t */

$$\textstyle\sum_{i=1}^{N}(\delta^2_{i,t} + \sum_{t'=t+1}^{T}\delta^3_{i,t,t'} + \sum_{t'=t+1}^{T-1}\delta^4_{i,t,t'}) \le 1, \quad t=1,2,..,T\text{-}2. \tag{3.2}$$

$$\textstyle\sum_{i=1}^{N}(\delta^2_{i,T-1} + \delta^3_{i,T-1,T}) \le 1. \tag{3.3}$$

$$(\delta^1_{i,t} + \delta^2_{i,t} + \sum_{t'=t+1}^{T} \delta^3_{i,t,t'} + \sum_{t'=t+1}^{T-1} \delta^4_{i,t,t'}) \leq 1 \quad \forall i \text{ and } t=1,2,..,T\text{-}2. \tag{3.4}$$

$$(\delta^1_{i,T-1} + \delta^2_{i,T-1} + \delta^3_{i,T-1,T}) \leq 1 \quad \forall i. \tag{3.5}$$

$$\sum_{i=1}^{N} (\Delta^1_{i,1,2} + \Delta^2_{i,1,2} + \Delta^3_{i,1,2,2}) \leq 1. \tag{3.6}$$

$$\sum_{i=1}^{N} \sum_{t''=1}^{t-1} (\Delta^1_{i,t'',t} + \Delta^2_{i,t'',t})$$
$$+ \sum_{i=1}^{N} \sum_{t'''=t''+1}^{t} \sum_{t''=1}^{t-1} \Delta^3_{i,t'',t''',t}$$
$$+ \sum_{i=1}^{N} \sum_{t'''=t''+1}^{t-1} \sum_{t''=1}^{t-2} \Delta^4_{i,t'',t''',t} \leq 1, \quad t=3,4,..,T. \tag{3.7}$$

/* Constraints (3.8)–(3.14) capture a possible complete setup within period t, with the production starting in period t */

$$\Delta^1_{i,t,t} = \delta^1_{i,t} \quad \forall i \text{ and } \forall t. \tag{3.8}$$

$$\Delta^1_{i,t,t'} \geq \Delta^1_{i,t,t'+1} \quad \forall i,\ t=1,2,..,T\text{-}1 \text{ and } t'=t,t+1,..,T\text{-}1. \tag{3.9}$$

$$X^1_{i,t,t'} \geq \mathcal{E}_d - \frac{C_{t'}}{a_i}(1 - \Delta^1_{i,t,t'}) \quad \forall i,\ \forall t \text{ and } t'=t,t+1,..,T. \tag{3.10}$$

$$a_i\, X^1_{i,t,t'} \leq C_{t'}\, \Delta^1_{i,t,t'} \quad \forall i,\ \forall t \text{ and } t'=t,t+1,..,T. \tag{3.11}$$

$$s^1_{i,t} = ST_i\, \delta^1_{i,t} \quad \forall i \text{ and } \forall t. \tag{3.12}$$

$$s^1_{i,t} \leq (C_t - \mathcal{E}) + C_t(1 - \delta^1_{i,t}) \quad \forall i \text{ and } \forall t. \tag{3.13}$$

$$\sum_{i'=1}^{N} (\delta^1_{i',t'-1} + \delta^2_{i',t'-1}$$
$$+ \sum_{t'''=t''+1}^{T} \sum_{t''=t}^{t'-1} \delta^3_{i',t'',t'''}$$
$$+ \sum_{t'''=t''+1}^{T-1} \sum_{t''=t}^{t'-1} \delta^4_{i',t'',t'''}) \leq N(1 - \Delta^1_{i,t,t'}) \quad \forall i,\ t=1,2,..,T\text{-}2 \text{ and } t'=t+2,t+3,..,T. \tag{3.14}$$

/* Constraints (3.15)–(3.21) correspond to a possible end-of-period setup in period t, with the production starting in period $t+1$ */

$$\Delta^2_{i,t,t'} = \delta^2_{i,t} \quad \forall i,\ t=1,2,..,T\text{-}1 \text{ and } t'=t+1. \tag{3.15}$$

$$\Delta^2_{i,t,t'} \geq \Delta^2_{i,t,t'+1} \quad \forall i,\ t=1,2,..,T\text{-}2 \text{ and } t'=t+1,t+2,..,T\text{-}1. \tag{3.16}$$

$$X^2_{i,t,t'} \geq \mathcal{E}_d - \frac{C_{t'}}{a_i}(1 - \Delta^2_{i,t,t'}) \quad \forall i,\ t=1,2,..,T\text{-}1 \text{ and } t'=t+1,t+2,..,T. \tag{3.17}$$

$$a_i\, X^2_{i,t,t'} \leq C_{t'}\, \Delta^2_{i,t,t'} \qquad \forall i, t=1,2,..,T\text{-}1 \text{ and } t'=t+1,t+2,..,T. \tag{3.18}$$

$$s^2_{i,t} = ST_i\, \delta^2_{i,t} \qquad \forall i \text{ and } t=1,2,..,T\text{-}1. \tag{3.19}$$

$$s^2_{i,t} \leq C_t + C_t(1 - \delta^2_{i,t}) \qquad \forall i \text{ and } t=1,2,..,T\text{-}1. \tag{3.20}$$

$$\sum\nolimits_{i'=1}^{N}(\delta^1_{i',t'-1} + \delta^2_{i',t'-1}$$
$$+ \sum\nolimits_{t'''=t''+1}^{T} \sum\nolimits_{t''=t}^{t'-1} \delta^3_{i',t'',t'''}$$
$$+ \sum\nolimits_{t'''=t''+1}^{T-1} \sum\nolimits_{t''=t}^{t'-1} \delta^4_{i',t'',t'''}) \leq N(1 - \Delta^2_{i,t,t'}) \qquad \forall i, t=1,2,..,T\text{-}2 \text{ and } t'=t+2,t+3,..,T. \tag{3.21}$$

/* Constraints (3.22)–(3.33) represent a possible setup crossover across a number of periods with the production starting in the same period where the setup ends */

$$\Delta^3_{i,t,t',t'} = \delta^3_{i,t,t'} \qquad \forall i, t=1,2,..,T\text{-}1 \text{ and } t'=t+1,t+2,..,T. \tag{3.22}$$

$$\Delta^3_{i,t,t',t''} \geq \Delta^3_{i,t,t',t''+1} \qquad \forall i, t=1,2,..,T\text{-}2, t'=t+1,t+2,..,T\text{-}1 \text{ and } t''=t',t'+1,..,T\text{-}1. \tag{3.23}$$

$$X^3_{i,t,t',t''} \geq \mathcal{E}_d - \tfrac{C_{t''}}{a_i}(1 - \Delta^3_{i,t,t',t''}) \qquad \forall i, t=1,2,..,T\text{-}1, t'=t+1,t+2,..,T \text{ and } t''=t',t'+1,..,T. \tag{3.24}$$

$$a_i\, X^3_{i,t,t',t''} \leq C_{t''}\, \Delta^3_{i,t,t',t''} \qquad \forall i, t=1,2,..,T\text{-}1, t'=t+1,t+2,..,T \text{ and } t''=t',t'+1,..,T. \tag{3.25}$$

$$s^3_{i,t,t',t''} \geq \mathcal{E} - C_{t''}(1 - \delta^3_{i,t,t'}) \qquad \forall i, t=1,2,..,T\text{-}1, t'=t+1,t+2,..,T \text{ and } t''=t,t+1,..,t'. \tag{3.26}$$

$$s^3_{i,t,t',t''} \leq C_{t''}\, \delta^3_{i,t,t'} \qquad \forall i, t=1,2,..,T\text{-}1, t'=t+1,t+2,..,T \text{ and } t''=t,t+1,..,t'. \tag{3.27}$$

$$\sum\nolimits_{t''=t}^{t'} s^3_{i,t,t',t''} = ST_i\, \delta^3_{i,t,t'} \qquad \forall i, t=1,2,..,T\text{-}1 \text{ and } t'=t+1,t+2,..,T. \tag{3.28}$$

$$s^3_{i,t,t',t''} = C_{t''}\, \delta^3_{i,t,t'} \qquad \forall i, t=1,2,..,T\text{-}2, t'=t+2,t+3,..,T \text{ and } t''=t+1,t+2,..,t'\text{-}1. \tag{3.29}$$

$$\sum_{i'=1}^{N}(\sum_{t''=t+1}^{t'-1}\delta^1_{i',t''}+\sum_{t''=t+1}^{t'-1}\delta^2_{i',t''}$$

$$+\sum_{t''=t+1}^{t'-1}\delta^3_{i',t,t''}+\sum_{t''=t'+1}^{T}\delta^3_{i',t,t''}$$

$$+\sum_{t'''=t''+1}^{T}\sum_{t''=t+1}^{t'-1}\delta^3_{i',t'',t'''}$$

$$+\sum_{t'''=t''+1}^{T-1}\sum_{t''=t}^{t'-1}\delta^4_{i',t'',t'''})$$

$$\leq (N\times T)\times(1-\sum_{i=1}^{N}\delta^3_{i,t,t'}), \quad t=1,2,..,T\text{-}3 \text{ and } t'=t+2,t+3,..,T\text{-}1. \tag{3.30}$$

$$\sum_{i'=1}^{N}(\sum_{t''=t+1}^{t'-1}\delta^1_{i',t''}+\sum_{t''=t+1}^{t'-1}\delta^2_{i',t''}$$

$$+\sum_{t''=t+1}^{t'-1}\delta^3_{i',t,t''}+\sum_{t'''=t''+1}^{T}\sum_{t''=t+1}^{t'-1}\delta^3_{i',t'',t'''}$$

$$+\sum_{t'''=t''+1}^{T-1}\sum_{t''=t}^{t'-2}\delta^4_{i',t'',t'''})$$

$$\leq (N\times T)\times(1-\sum_{i=1}^{N}\delta^3_{i,t,t'}), \quad t=1,2,..,T\text{-}2 \text{ and } t'=T. \tag{3.31}$$

$$\sum_{i'=1}^{N}(\delta^1_{i',t''-1}+\delta^2_{i',t''-1}+\sum_{t'''=t''}^{T}\delta^3_{i',t''-1,t'''}$$

$$+\sum_{t'''=t''}^{T-1}\delta^4_{i',t''-1,t'''})\leq N(1-\Delta^3_{i,t,t',t''}) \quad \forall i,\ t=1,2,..,T\text{-}3,\ t'=t+1,t+2,..,T\text{-}2 \text{ and } t''=t'+1,t'+2,..,T\text{-}1. \tag{3.32}$$

$$\sum_{i'=1}^{N}(\delta^1_{i',t''-1}+\delta^2_{i',t''-1}+\sum_{t'''=t''}^{T}\delta^3_{i',t''-1,t'''})$$

$$\leq N(1-\Delta^3_{i,t,t',t''}) \quad \forall i,\ t=1,2,..,T\text{-}2,\ t'=t+1,t+2,..,T\text{-}1 \text{ and } t''=T. \tag{3.33}$$

/* Constraints (3.34)–(3.45) represent a possible setup crossover across a number of periods with the production starting in period $t'+1$ when the setup ends at the end of period t' */

$$\Delta^4_{i,t,t',t'+1}=\delta^4_{i,t,t'} \quad \forall i,\ t=1,2,..,T\text{-}2 \text{ and } t'=t+1,t+2,..,T\text{-}1. \tag{3.34}$$

$$\Delta^4_{i,t,t',t''}\geq\Delta^4_{i,t,t',t''+1} \quad \forall i,\ t=1,2,..,T\text{-}3,\ t'=t+1,t+2,..,T\text{-}2 \text{ and } t''=t'+1,t'+2,..,T\text{-}1. \tag{3.35}$$

$$X^4_{i,t,t',t''}\geq\mathcal{E}_d-\frac{C_{t''}}{a_i}(1-\Delta^4_{i,t,t',t''}) \quad \forall i,\ t=1,2,..,T\text{-}2,\ t'=t+1,t+2,..,T\text{-}1 \text{ and } t''=t'+1,t'+2,..,T. \tag{3.36}$$

$$a_i\,X^4_{i,t,t',t''}\leq C_{t''}\,\Delta^4_{i,t,t',t''} \quad \forall i,\ t=1,2,..,T\text{-}2,\ t'=t+1,t+2,..,T\text{-}1 \text{ and } t''=t'+1,t'+2,..,T. \tag{3.37}$$

$$s^4_{i,t,t',t''}\geq\mathcal{E}-C_{t''}(1-\delta^4_{i,t,t'}) \quad \forall i,\ t=1,2,..,T\text{-}2,\ t'=t+1,t+2,..,T\text{-}1 \text{ and } t''=t,t+1,..,t'. \tag{3.38}$$

$$s^4_{i,t,t',t''} \leq C_{t''}\, \delta^4_{i,t,t'} \qquad \forall i, t\text{=}1,2,..,T\text{-}2, t'\text{=}t\text{+}1,t\text{+}2,..,T\text{-}1 \text{ and } t''\text{=}t,t\text{+}1,..,t'. \tag{3.39}$$

$$\sum_{t''=t}^{t'} s^4_{i,t,t',t''} = ST_i\, \delta^4_{i,t,t'} \qquad \forall i, t\text{=}1,2,..,T\text{-}2 \text{ and } t'\text{=}t\text{+}1,t\text{+}2,..,T\text{-}1. \tag{3.40}$$

$$s^4_{i,t,t',t''} = C_{t''}\, \delta^4_{i,t,t'} \qquad \forall i, t\text{=}1,2,..,T\text{-}2, t'\text{=}t\text{+}1,t\text{+}2,..,T\text{-}1 \text{ and } t''\text{=}t\text{+}1,t\text{+}2,..,t'. \tag{3.41}$$

$$\sum_{i'=1}^{N}\Big(\sum_{t''=t+1}^{t'-1} \delta^1_{i',t''} + \sum_{t''=t+1}^{t'-1} \delta^2_{i',t''} + \sum_{t'''=t''+1}^{T} \sum_{t''=t}^{t'-1} \delta^3_{i',t'',t'''} + \sum_{t''=t+1}^{t'-1} \delta^4_{i',t,t''} + \sum_{t''=t'+1}^{T-1} \delta^4_{i',t,t''} + \sum_{t'''=t''+1}^{T-1} \sum_{t''=t+1}^{t'-1} \delta^4_{i',t'',t'''}\Big) \leq (N \times T) \times \Big(1 - \sum_{i=1}^{N} \delta^4_{i,t,t'}\Big), \qquad t\text{=}1,2,..,T\text{-}4 \text{ and } t'\text{=}t\text{+}2,t\text{+}3,..,T\text{-}2. \tag{3.42}$$

$$\sum_{i'=1}^{N}\Big(\sum_{t''=t+1}^{t'-1} \delta^1_{i',t''} + \sum_{t''=t+1}^{t'-1} \delta^2_{i',t''} + \sum_{t'''=t''+1}^{T} \sum_{t''=t}^{t'-1} \delta^3_{i',t'',t'''} + \sum_{t''=t+1}^{t'-1} \delta^4_{i',t,t''} + \sum_{t'''=t''+1}^{T-1} \sum_{t''=t+1}^{t'-1} \delta^4_{i',t'',t'''}\Big) \leq (N \times T) \times \Big(1 - \sum_{i=1}^{N} \delta^4_{i,t,t'}\Big), \qquad t\text{=}1,2,..,T\text{-}3 \text{ and } t'\text{=}T\text{-}1. \tag{3.43}$$

$$\sum_{i'=1}^{N}\Big(\delta^1_{i',t''-1} + \delta^2_{i',t''-1} + \sum_{t'''=t''}^{T} \delta^3_{i',t''-1,t'''} + \sum_{t'''=t''}^{T-1} \delta^4_{i',t''-1,t'''}\Big) \leq N(1 - \Delta^4_{i,t,t',t''}) \qquad \forall i, t\text{=}1,2,..,T\text{-}3, t'\text{=}t\text{+}1,t\text{+}2,..,T\text{-}2 \text{ and } t''\text{=}t'\text{+}1,t'\text{+}2,..,T\text{-}1. \tag{3.44}$$

$$\sum_{i'=1}^{N}\Big(\delta^1_{i',t''-1} + \delta^2_{i',t''-1} + \sum_{t'''=t''}^{T} \delta^3_{i',t''-1,t'''}\Big) \leq N(1 - \Delta^4_{i,t,t',t''}) \qquad \forall i, t\text{=}1,2,..,T\text{-}2, t'\text{=}t\text{+}1,t\text{+}2,..,T\text{-}1 \text{ and } t''\text{=}T. \tag{3.45}$$

/* Constraints (3.46)–(3.49) indicate that a production carryover of any product i can take place in period $t + 2 \leq t'' \leq T$, only if the sum of possible production times and setup times of all products in period $t'' - 1$ is equal to the capacity of period $t'' - 1$ */

$$\sum_{i=1}^{N}(a_i\,X^1_{i,t,t''-1}+a_i\,X^2_{i,t,t''-1}+a_i\,X^3_{i,t,t''-1,t''-1}$$
$$+\sum_{t'=t''-1}^{t''}s^3_{i,t,t',t''-1}+\sum_{t'=t''-1}^{t''}s^4_{i,t,t',t''-1})$$
$$\leq C_{t''-1}+((C_{t''-1}+1)\times(1-\sum_{i=1}^{N}(\Delta^1_{i,t,t''}$$
$$+\Delta^2_{i,t,t''}+\sum_{t'=t+1}^{t''}\Delta^3_{i,t,t',t''}$$
$$+\sum_{t'=t+1}^{t''-1}\Delta^4_{i,t,t',t''}))),\qquad t=1,2,..,T\text{-}2 \text{ and } t''=t+2. \tag{3.46}$$

$$\sum_{i=1}^{N}(a_i\,X^1_{i,t,t''-1}+a_i\,X^2_{i,t,t''-1}$$
$$+a_i\sum_{t'=t+1}^{t''-1}X^3_{i,t,t',t''-1}+a_i\sum_{t'=t+1}^{t''-2}X^4_{i,t,t',t''-1}$$
$$+\sum_{t'=t''-1}^{t''}s^3_{i,t,t',t''-1}$$
$$+\sum_{t'=t''-1}^{t''}s^4_{i,t,t',t''-1})\leq C_{t''-1}+((C_{t''-1}+1)\times$$
$$(1-\sum_{i=1}^{N}(\Delta^1_{i,t,t''}+\Delta^2_{i,t,t''}$$
$$+\sum_{t'=t+1}^{t''}\Delta^3_{i,t,t',t''}+\sum_{t'=t+1}^{t''-1}\Delta^4_{i,t,t',t''}))),\qquad t=1,2,..,T\text{-}3 \text{ and } t''=t+3,t+4,..,T. \tag{3.47}$$

$$\sum_{i=1}^{N}(a_i\,X^1_{i,t,t''-1}+a_i\,X^2_{i,t,t''-1}+a_i\,X^3_{i,t,t''-1,t''-1}$$
$$+\sum_{t'=t''-1}^{t''}s^3_{i,t,t',t''-1}+\sum_{t'=t''-1}^{t''}s^4_{i,t,t',t''-1})$$
$$\geq C_{t''-1}-((C_{t''-1}+1)\times(1-\sum_{i=1}^{N}(\Delta^1_{i,t,t''}$$
$$+\Delta^2_{i,t,t''}+\sum_{t'=t+1}^{t''}\Delta^3_{i,t,t',t''}$$
$$+\sum_{t'=t+1}^{t''-1}\Delta^4_{i,t,t',t''}))),\qquad t=1,2,..,T\text{-}2 \text{ and } t''=t+2. \tag{3.48}$$

$$\sum_{i=1}^{N}(a_i\,X^1_{i,t,t''-1}+a_i\,X^2_{i,t,t''-1}$$
$$+a_i\sum_{t'=t+1}^{t''-1}X^3_{i,t,t',t''-1}+a_i\sum_{t'=t+1}^{t''-2}X^4_{i,t,t',t''-1}$$
$$+\sum_{t'=t''-1}^{t''}s^3_{i,t,t',t''-1}+\sum_{t'=t''}^{t''}s^4_{i,t,t',t''-1})$$
$$\geq C_{t''-1}-((C_{t''-1}+1)\times$$
$$(1-\sum_{i=1}^{N}(\Delta^1_{i,t,t''}+\Delta^2_{i,t,t''}$$
$$+\sum_{t'=t+1}^{t''}\Delta^3_{i,t,t',t''}+\sum_{t'=t+1}^{t''-1}\Delta^4_{i,t,t',t''}))),\qquad t=1,2,..,T\text{-}3 \text{ and } t''=t+3,t+4,..,T. \tag{3.49}$$

/* Constraints (3.50)–(3.54) represent the feasibility of production and setup with respect to capacity availability */

$$\sum_{i=1}^{N}(s_{i,1}^{1}+s_{i,1}^{2}+\sum_{t''=2}^{T}s_{i,1,t'',1}^{3}+\sum_{t''=2}^{T-1}s_{i,1,t'',1}^{4}$$

$$+a_i\,X_{i,1,1}^{1})\leq C_1. \tag{3.50}$$

$$\sum_{i=1}^{N}(s_{i,2}^{1}+s_{i,2}^{2}+\sum_{t''=2}^{T}s_{i,1,t'',2}^{3}+\sum_{t''=3}^{T}s_{i,2,t'',2}^{3}$$

$$+\sum_{t''=2}^{T-1}s_{i,1,t'',2}^{4}+\sum_{t''=3}^{T-1}s_{i,2,t'',2}^{4}$$

$$+\sum_{t''=1}^{2}a_i\,X_{i,t'',2}^{1}+a_i\,X_{i,1,2}^{2}+a_i\,X_{i,1,2,2}^{3})\leq C_2. \tag{3.51}$$

$$\sum_{i=1}^{N}(s_{i,t}^{1}+s_{i,t}^{2}+\sum_{t''=t}^{T}\sum_{t'=1}^{t-1}s_{i,t',t'',t}^{3}$$

$$+\sum_{t''=t+1}^{T}s_{i,t,t'',t}^{3}+\sum_{t''=t}^{T-1}\sum_{t'=1}^{t-1}s_{i,t',t'',t}^{4}$$

$$+\sum_{t''=t+1}^{T-1}s_{i,t,t'',t}^{4})+\sum_{i=1}^{N}\sum_{t''=1}^{t}a_i\,X_{i,t'',t}^{1}$$

$$+\sum_{i=1}^{N}\sum_{t''=1}^{t-1}a_i\,X_{i,t'',t}^{2}$$

$$+\sum_{i=1}^{N}\sum_{t'''=t''+1}^{t}\sum_{t''=1}^{t-1}a_i\,X_{i,t'',t''',t}^{3}$$

$$+\sum_{i=1}^{N}\sum_{t'''=t''+1}^{t-1}\sum_{t''=1}^{t-2}a_i\,X_{i,t'',t''',t}^{4}\leq C_t, \qquad t=3,4,..,T\text{-}1. \tag{3.52}$$

$$\sum_{i=1}^{N}(s_{i,T}^{1}+\sum_{t'=1}^{T-1}s_{i,t',T,T}^{3})$$

$$+\sum_{i=1}^{N}\sum_{t''=1}^{T}a_i\,X_{i,t'',T}^{1}$$

$$+\sum_{i=1}^{N}\sum_{t''=1}^{T-1}a_i\,X_{i,t'',T}^{2}$$

$$+\sum_{i=1}^{N}\sum_{t'''=t''+1}^{T}\sum_{t''=1}^{T-1}a_i\,X_{i,t'',t''',T}^{3}$$

$$+\sum_{i=1}^{N}\sum_{t'''=t''+1}^{T-1}\sum_{t''=1}^{T-2}a_i\,X_{i,t'',t''',T}^{4}\leq C_T. \tag{3.53}$$

$$s_{i,t,T,t'}^{4}=0 \qquad \forall i,\ t=1,2,..,T \text{ and } t'=t,t+1,..,T. \tag{3.54}$$

/* Constraints (3.55)–(3.57) represent the inventory balance constraints; constraint (4.58) assumes that there is neither inventory nor backorder of any product at the end of the planning horizon */

$$I_{i,1}-B_{i,1}=X_{i,1,1}^{1}-d_{i,1} \qquad \forall i. \tag{3.55}$$

$$I_{i,2}-B_{i,2}=X_{i,1,1}^{1}+X_{i,1,2}^{1}+X_{i,2,2}^{1}+X_{i,1,2}^{2}$$

$$+X_{i,1,2,2}^{3}-d_{i,1}-d_{i,2} \qquad \forall i. \tag{3.56}$$

$$I_{i,t} - B_{i,t} = \sum_{t'''=1}^{t} \sum_{t''=1}^{t'''} X^1_{i,t'',t'''}$$

$$+ \sum_{t'''=2}^{t} \sum_{t''=1}^{t'''-1} X^2_{i,t'',t'''}$$

$$+ \sum_{t'=t'''}^{t} \sum_{t'''=2}^{t} \sum_{t''=1}^{t'''-1} X^3_{i,t'',t''',t'}$$

$$+ \sum_{t'=t'''+1}^{t} \sum_{t'''=2}^{t-1} \sum_{t''=1}^{t'''-1} X^4_{i,t'',t''',t'}$$

$$- \sum_{t'''=1}^{t} d_{i,t'''} \qquad \forall i \text{ and } t=3,4,..,T. \tag{3.57}$$

$$I_{i,T} = B_{i,T} = 0 \qquad \forall i. \tag{3.58}$$

$\delta^1_{i,t}$, $\delta^2_{i,t}$, $\delta^3_{i,t,t'}$, $\delta^4_{i,t,t'}$, $\Delta^1_{i,t,t'}$, $\Delta^2_{i,t,t'}$, $\Delta^3_{i,t,t',t''}$ and $\Delta^4_{i,t,t',t''}$ are binary, and all other variables are ≥ 0.

Note: $\delta^1_{i,0} = \delta^2_{i,0} = 0 \quad \forall i$ and $\delta^3_{i,0,t} = \delta^4_{i,0,t} = 0 \quad \forall i$ and $\forall t$.

Also in Eq. (3.58), $B_{i,T} = 0 \ \forall i$ exists when backorders are not supposed to be present at the end of a planning horizon (i.e. here the demand for all products should be satisfied within period T). In contrary, the equation must be removed when demand for all products need not be satisfied within period T.

The objective function shown in Eq. (3.1) is the minimization of the time-dependent setup cost, time-dependent holding cost and backorder cost of all products across all time periods. Constraints (3.2)–(3.3) indicate that there can be either an end-of-period setup or a setup crossover for at most one product in every period t. Constraints (3.4)–(3.5) indicate that for every product i, there can be at most one of the setup-types in each period t. Constraints (3.6)–(3.7) indicate the conditions for a production carryover to occur between adjacent periods. Constraints (3.8), (3.15), (3.22) and (3.34) indicate that when there is a setup of product i in period t, a corresponding production is carried out. Constraints (3.9), (3.16), (3.23) and (3.35) ensure that a production carryover in period $t' + 1$ is not possible without a corresponding production in period t'. Constraints (3.10), (3.17), (3.24) and (3.36) address the condition that a production exists corresponding to the indicator variable (i.e. when the indicator variable takes the value 1, at least a small $\mathcal{E}_d$ unit of the corresponding product should be produced). Constraints (3.11), (3.18), (3.25) and (3.37) ensure that the production quantity of a product in period t does not exceed the capacity of the same period (measured in time units). Constraints (3.12), (3.19), (3.28) and (3.40) ensure that when the setup of a product i is initiated in period t (i.e. when $\delta^1_{i,t}$ or $\delta^2_{i,t}$ or $\delta^3_{i,t,t'}$ or $\delta^4_{i,t,t'}$ exists), the total setup time ST_i required by product i for the setup is consumed. Constraint (3.13) shows that a product i which is set up in period t using $\delta^1_{i,t}$ can have a setup time less than or equal to C_t - $\mathcal{E}$ and should not fill the entire capacity of that period. Constraint (3.20) shows that a product i which is set up in period t using $\delta^2_{i,t}$ can have a setup time less than or equal to C_t, or is allowed to fill the entire capacity of that period. Constraints (3.26) and (3.38) indicate that when products having long setup times are set up using $\delta^3_{i,t,t'}$ and $\delta^4_{i,t,t'}$, then from period t where the setup starts to period t' where the setup ends, at least an $\mathcal{E}$ (non-zero) unit of setup time should be present in those periods. Constraints (3.27) and (3.39) indicate that the setup times of a product setup using $\delta^3_{i,t,t'}$ and $\delta^4_{i,t,t'}$ cannot take a value greater than the capacity of the period in which the setup is carried over. Constraints (3.14), (3.21), (3.32), (3.33), (3.44) and (3.45) indicate that a production can be carried over to a period from the previous period without any interruption, only if there is no intermittent setup of any other product. Constraints (3.29) and (3.41) show the condition for a setup crossover to occur from one period to the next. Constraints (3.30), (3.31), (3.42) and (3.43) indicate that when a setup is carried over across periods, no setup of any other product should exist in the intermittent periods. Constraints (3.46)–(3.49) indicate the conditions for an uninterrupted production to occur (production without idle time introduced in between), i.e., the constraints indicate that the production carryover of product i can take place in period t' due to the setup of product i in period t, only if the sum of the possible setup times and production times of all products is equal to the capacity of the previous period $t'-1$. Constraints (3.50)–(3.54) denote the capacity constraints, with the consideration of possible setup times and production times of all products. Constraints (3.55)–(3.57) represent the inventory balance constraints. Constraint (3.58) assumes that there is neither inventory nor backorder of any product at the end of the planning horizon. It is to be noted that the parameters $\mathcal{E}_d$ and $\mathcal{E}$ are set to a small positive real value.

3.3.4 Method of Tracking Setups in MM1:CLSP-PCSC

A brief explanation on the principle of tracking the setups in MM1:CLSP-PCSC especially for multi-period setups (within a time period/across time periods), the corresponding setup times, and production periods is given in this section.

In MM1:CLSP-PCSC, the multi-period setups are tracked with the help of two different variables, $\delta^3_{i,t,t'}$ and $\delta^4_{i,t,t'}$. In both these variables, t represents the period in which setup starts and t' represents the period in which setup ends. $\delta^3_{i,t,t'}$ is used to track the setup which starts in period t and runs across one or more periods, and gets completed before the end of a future period t'. $\delta^4_{i,t,t'}$ tracks the multi-period setup which starts in period t, runs across one or more periods and gets completed at the end of a future period t'.

The variables $s^3_{i,t,t',t''}$ and $s^4_{i,t,t',t''}$ help to track the setup times. Here t represents the period when the setup starts; t' represents a future period when the setup gets completed and t'' represents the period in which the setup time is present ($t \leq t'' \leq t'$).

The variables $\Delta^3_{i,t,t',t''}$ and $\Delta^4_{i,t,t',t''}$ represent the production variables of the multi-period setup, corresponding to $\delta^3_{i,t,t'}$ and $\delta^4_{i,t,t'}$, respectively. Here t represents the period in which the setup starts; t' represents a future period in which the setup ends and t'' represents the period in which the production occurs.

3.4 Special Cases of CLSP-PCSC with Respect to MM1:CLSP-PCSC

Two special cases of CLSP-PCSC with respect to MM1:CLSP-PCSC are presented in this section. They are: (i) setup cost of a product calculated with respect to the period of its setup completion and (ii) setup cost and holding cost of a product are time independent.

3.4.1 Setup Cost of a Product Calculated with Respect to the Period of Its Setup Completion

When the cost of setup for a product is computed with respect to the period where the setup ends, in case of MM1:CLSP-PCSC the objective function is modified.

The objective function of Mathematical Model 1 is:

$$\text{Min } Z = \sum_{i=1}^{N}\sum_{t=1}^{T} SC_{i,t}\,\delta^1_{i,t} + \sum_{i=1}^{N}\sum_{t=1}^{T-1} SC_{i,t}\,\delta^2_{i,t} + \sum_{i=1}^{N}\sum_{t=1}^{T-1}\sum_{t'=t+1}^{T} SC'_{i,t'}\,\delta^3_{i,t,t'}$$

$$+\sum_{i=1}^{N}\sum_{t=1}^{T-2}\sum_{t'=t+1}^{T-1} SC'_{i,t'}\,\delta^4_{i,t,t'} + \sum_{i=1}^{N}\sum_{t=1}^{T} h_{i,t}\,I_{i,t} + \sum_{i=1}^{N}\sum_{t=1}^{T} b_i\,B_{i,t} \tag{3.59}$$

Subject to constraints (3.2)–(3.58).
Where $SC'_{i,t'}$—setup cost for product i with its setup getting completed in period t'; this cost is incurred only once as a fixed cost computed with respect to the period of its setup completion (*Note:* $SC_{i,t} = SC'_{i,t}$ with respect to $\delta^1_{i,t}$ and $\delta^2_{i,t}$ for this special case of CLSP-PCSC).

3.4.2 Setup Cost and Holding Cost of a Product Being Time Independent

When the cost for setup is product dependent and time independent, the objective function for MM1:CLSP-PCSC is modified as follows:

$$\text{Min } Z = \sum_{i=1}^{N}\sum_{t=1}^{T} SC'_i\,\delta^1_{i,t} + \sum_{i=1}^{N}\sum_{t=1}^{T-1} SC'_i\,\delta^2_{i,t} + \sum_{i=1}^{N}\sum_{t=1}^{T-1}\sum_{t'=t+1}^{T} SC'_i\,\delta^3_{i,t,t'}$$

$$+\sum_{i=1}^{N}\sum_{t=1}^{T-2}\sum_{t'=t+1}^{T-1} SC'_i\,\delta^4_{i,t,t'} + \sum_{i=1}^{N}\sum_{t=1}^{T} h'_i\,I_{i,t} + \sum_{i=1}^{N}\sum_{t=1}^{T} b_i\,B_{i,t} \tag{3.60}$$

subject to constraints (3.2)–(3.58).

Where SC'_i—setup cost for product i, and h'_i—holding cost for product i.

3.5 Numerical Illustrations and Discussion with Respect to MM1:CLSP-PCSC

Let us consider a sample problem instance shown in Table 3.1, with the demand data ($d_{i,t}$) given for five products across sixteen time periods. The data for the time-dependent setup costs ($SC_{i,t}$) and holding costs ($h_{i,t}$) are given in the table. The values of the setup time (ST_i), number of time units required for producing one unit of product i (a_i) and backorder costs (b_i) for all products are also given in the table. All these values provided in Table 3.1 are used to illustrate the proposed MM1:CLSP-PCSC which addresses the case, 'Setup cost of a product calculated with respect to the period its setup initiation', and other special cases, namely 'Setup cost of a product calculated with respect to the period of its setup completion' and 'Situations present in industries manufacturing discrete products'. Values for time-independent setup costs for the five products considered are $SC'_i = \{50, 10, 30, 20, 40\}$, $i = 1, 2, 3, 4, 5$, and time-independent holding costs for the five products are $h'_i = \{2, 3, 1, 1, 2\}$, $i = 1, 2, 3, 4, 5$. These values along with the values $d_{i,t}$, ST_i, b_i and a_i provided in Table 3.1 are used to illustrate a special case, namely 'Setup cost and holding cost of a product are time independent'. The example depicts identical capacity across all time periods for the sake of illustration. However, our model allows for different capacities across different time periods due to the use of C_t (to describe variable capacities across different time periods), without loss of generality.

3.5.1 Setup Cost of a Product Calculated with Respect to the Period of Its Setup Initiation

The proposed Mathematical Model 1 presented in Sect. 3.3.3 of this book considers the presence of time-dependent setup costs and holding costs with the setup cost of a product calculated with respect to the period of its setup initiation. When the MM1:CLSP-PCSC is executed for the given sample problem instance given in Table 3.1 considering time-dependent setup costs and holding costs, a solution (see Table 3.2) with its corresponding Gantt chart shown in Fig. 3.3 is obtained, with the value of the objective function equal to 200 mu (monetary units). Product 4 is produced with two setups (see periods 8, 9 and 10) with a break in production (maintaining the feasibility condition with respect to production occurring immediately after setup and no break in production carryover). This is so because the setup cost of product 4 is smaller than its holding cost up to that period. For the same problem instance given in Table 3.1, the setup cost of product 4 is increased from 60 mu to 120 mu in period 9, and increased from 20 mu to 120 mu in period 10. Now it is observed that product 4 is continuously produced in periods 8, 9 and 10 with a single setup (see Fig. 3.4). This is so because the setup cost of product 4 is greater than its holding cost. The objective function value obtained in this case (when MM1:CLSP-PCSC is executed) is equal to $Z = 270$ mu. In both cases (Figs. 3.3 and 3.4) it is observed that the production of all products is carried out immediately after setup and there is no interruption or break in production carryover. Hence the solution obtained is feasible and optimal for the problem statement considered with respect to process industries. Through these observations it can be understood that the proposed mathematical model addresses situations where an uninterrupted production has to take place (see Fig. 3.4), and if there is any discontinuity or break in production across periods (due to a lower setup cost compared to the holding cost) a new setup can be done for the same product for any subsequent production (see Fig. 3.3).

3.5.2 Setup Cost of a Product Calculated with Respect to the Period of Its Setup Completion

This special case of MM1:CLSP-PCSC where the setup cost is calculated in period t' (period when the setup ends) is illustrated with the help of the numerical example shown in Table 3.1. Here the products are assumed to have time-dependent and product-dependent setup costs ($SC'_{i,t'}$) and holding costs ($h_{i,t}$), whose values are given in Table 3.1. This special case of the mathematical model generates a Gantt chart as shown in Fig. 3.5. The value of the objective function generated for this case is Z = 190 mu.

3.5.3 Setup Cost and Holding Cost of a Product Being Time Independent

The time-independent cost structure (i.e. SC'_i and h'_i whose values are provided in this section) is considered along with the demand and backorder cost data (i.e. $d_{i,t}$ and b_i) shown in Table 3.1 to illustrate this special case. This special case of MM1:CLSP-PCSC generates a Gantt chart as shown in Fig. 3.6. The value of the objective function generated for this case is Z = 220 mu.

3.5.4 Observations from an Existing Model

The model developed by Belo-Filho et al. (2013) also attempts to address situations in process industries. For the purpose of illustrating the model by Belo-Filho et al. (2013) in terms of not completely addressing the characteristics of process industries (i.e. not addressing the cases of production of a product starting immediately after setup, and uninterrupted production carryover across periods), a numerical illustration with the following data is presented (i.e. $h_{i,t} = h_{i,t+1} = h'_i, \ \ t = 1, 2, \ldots, T-1$ and $\forall i$; and $SC_{i,t} = SC_{i,t+1} = SC'_i, \ t = 1, 2, \ldots, T-1$ and $\forall i$). It means that time invariant holding cost rate and setup cost for a given product across all time periods are present. Therefore, in the place of values of $h_{i,t}$ and $SC_{i,t}$ given in Table 3.1, the following values are set: $h'_i = \{2, 3, 1, 1, 2\}, \quad i = 1, 2, 3, 4, 5$; and $SC'_i = \{50, 10, 30, 20, 40\}, \quad i = 1, 2, 3, 4, 5$ (*Note:* The same values are used to illustrate the special case, namely 'Setup cost and holding cost of a product are time independent'). Here, h'_i denotes the time-independent holding cost for product i (mu/period/unit product carried over), and SC'_i denotes the time-independent setup cost for product i (mu/setup). When the model developed by Belo-Filho et al. (2013) (their *First proposed formulation*) is executed for this example (i.e. considering the product-dependent and time-independent setup costs and holding costs) the model does not provide a feasible and optimal solution when programmed as given in their paper. This is so because the first setup with respect to period 1 does not get generated while executing their formulation and in turn results in a wrong value of Z. When the constraint in their *First proposed formulation*, namely $\sum_{i=1}^{N} \alpha_{i,t} = 1$, is executed with $t = 2, 3, .., T$ instead of $\forall t$ as given in their paper, a feasible and an optimal solution is obtained. From this solution, it is observed that there is no production immediately after a setup (see periods 3, 11 and 15) and there is a break in the production carryover across periods (see periods 3, 4, 5, 8, 9 and 10 in Fig. 3.7). The value of Z equals 200 mu. The solution of the model by Belo-Filho et al. is shown in Table 3.3 with its corresponding Gantt chart shown in Fig. 3.7.

Table 3.1: A sample problem instance

Product	Setup time (ST_i): (time units)	Backorder cost (b_i): (mu/period/ unit product backordered)	Number of time units required for producing one unit of product i (a_i): (time units/unit of product i)
1	10	20	1
2	220	20	1
3	200	40	1
4	10	30	1
5	100	30	1

Product \ Period	1	2	3	4	5	6	7	8	9	10	11	12	13	14	15	16
Demand ($d_{i,t}$)																
1	60	0	0	0	0	0	0	0	0	0	0	0	0	0	0	0
2	0	0	0	90	20	0	0	0	0	0	0	0	0	0	0	0
3	0	0	0	0	0	0	45	0	0	0	0	0	0	0	0	100
4	0	0	0	0	0	0	0	80	70	60	0	0	50	0	0	0
5	0	0	0	0	0	0	0	0	0	0	0	100	10	0	0	0
Setup cost ($SC_{i,t}$): (mu/setup)																
1	20	30	10	30	50	20	40	20	40	30	20	10	20	40	30	20
2	40	20	40	30	20	10	10	20	10	30	40	20	10	40	60	50
3	30	50	60	40	20	40	30	60	40	20	10	30	40	20	10	20
4	20	40	30	20	10	30	60	30	60	20	50	20	10	30	60	20
5	60	30	20	20	30	10	30	30	20	60	40	40	50	20	10	30
Holding cost ($h_{i,t}$): (mu/period/unit product carried over)																
1	1	1	1	1	1	2	2	2	2	3	3	3	3	3	3	3
2	2	2	2	2	1	1	1	1	1	2	2	4	4	4	4	4
3	4	4	4	4	3	3	3	3	2	2	2	2	1	1	1	1
4	3	3	3	2	2	2	1	1	3	1	1	2	2	2	3	3
5	4	4	4	1	1	1	3	3	3	4	4	4	2	2	2	2
Capacity (C_t): time units	100	100	100	100	100	100	100	100	100	100	100	100	100	100	100	100

Table 3.2: Solution generated by the proposed mathematical model MM1:CLSP-PCSC (corresponding terms in MM1:CLSP-PCSC are used here) for the data provided in Table 3.1 with the corresponding Gantt chart provided in Fig. 3.3

$t = 1$	$t = 2$	$t = 3$	$t = 4$	$t = 5$	$t = 6$	$t = 7$	$t = 8$
$\delta^1_{1,1} = 1$ $\delta^3_{2,1,4} = 1$ $\Delta^1_{1,1,1} = 1$ $X^1_{1,1,1} = 60$ $s^1_{1,1} = 10$ $s^3_{2,1,4,1} = 10$	$s^3_{2,1,4,2} = 100$	$s^3_{2,1,4,3} = 100$	$\Delta^3_{2,1,4,4} = 1$ $X^3_{2,1,4,4} = 90$ $s^3_{2,1,4,4} = 10$	$\delta^3_{3,5,7} = 1$ $\Delta^3_{2,1,4,5} = 1$ $X^3_{2,1,4,5} = 20$ $s^3_{3,5,7,5} = 80$	$s^3_{3,5,7,6} = 100$	$\Delta^3_{3,5,7,7} = 1$ $X^3_{3,5,7,7} = 45$ $s^3_{3,5,7,7} = 20$	$\delta^1_{4,8} = 1$ $\Delta^1_{4,8,8} = 1$ $X^1_{4,8,8} = 80$ $s^1_{4,8} = 10$
$t = 9$	$t = 10$	$t = 11$	$t = 12$	$t = 13$	$t = 14$	$t = 15$	$t = 16$
$\Delta^1_{4,8,9} = 1$ $X^1_{4,8,9} = 70$	$\delta^1_{4,10} = 1$ $\Delta^1_{4,10,10} = 1$ $X^1_{4,10,10} = 60$ $s^1_{4,10} = 10$	$\delta^2_{5,11} = 1$ $s^2_{5,11} = 100$	$\Delta^2_{5,11,12} = 1$ $X^2_{5,11,12} = 100$	$\delta^1_{4,13} = 1$ $\Delta^1_{4,13,13} = 1$ $\Delta^2_{5,11,13} = 1$ $X^1_{4,13,13} = 50$ $X^2_{5,11,13} = 10$ $s^1_{4,13} = 10$	$\delta^4_{3,14,15} = 1$ $s^4_{3,14,15,14} = 100$	$s^4_{3,14,15,15} = 100$	$\Delta^4_{3,14,15,16} = 1$ $X^4_{3,14,15,16} = 100$

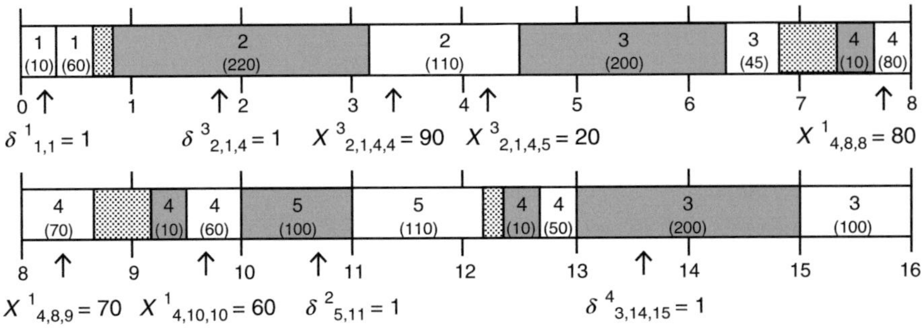

Figure 3.3: Gantt chart obtained by the proposed mathematical model MM1:CLSP-PCSC (for the data given in Table 3.1 and the solution provided in Table 3.2) when the setup costs and holding costs are time dependent and the setup cost of a product is calculated with respect to the period of its setup initiation; $Z = 200$ mu

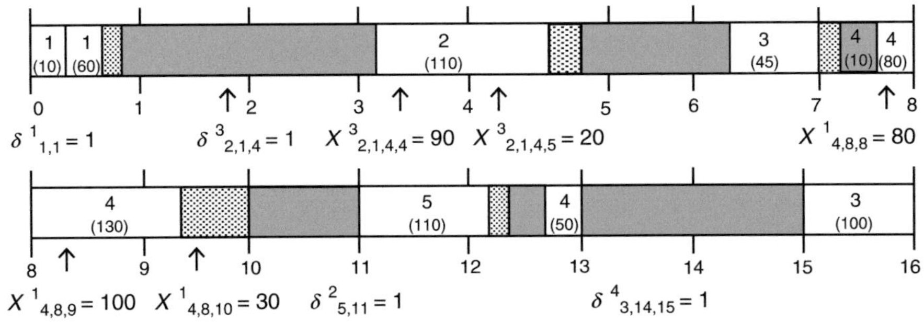

Figure 3.4: Gantt chart for the solution obtained by the proposed mathematical model MM1:CLSP-PCSC when the setup costs and holding costs are time dependent (for the data given in Table 3.1) and the setup cost of a product is calculated with respect to the period of its setup initiation. Here the setup cost for product 4 is increased from 60 mu to 120 mu in period 9, and from 20 mu to 120 mu in period 10; $Z =$ 270 mu

Remark 1 (with Respect to Figs. 3.3–3.7): Time period is denoted along the X-axis. The shaded region denotes the setup of the product, indicated inside the shaded region by 1, 2 and so on, with the corresponding setup time shown within the parenthesis. The non-shaded region denotes the production of the product following its setup, and the production time is indicated within the parenthesis. Idle time of a machine is denoted with dots inside. Values of some variables are shown underneath, over periods 1, 2, ..., 16 in the figure, for the sake of understanding.

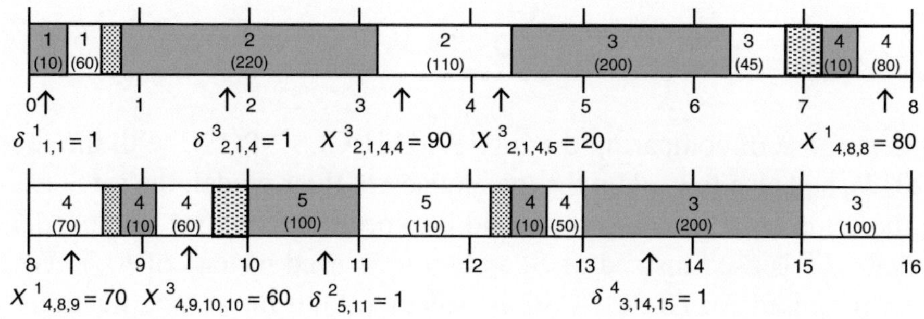

Figure 3.5: Gantt chart obtained by the proposed model MM1:CLSP-PCSC (for the data given in Table 3.1) when the setup costs and holding costs are time dependent and the setup cost of a product is calculated with respect to the period of its setup completion; $Z = 190$ mu

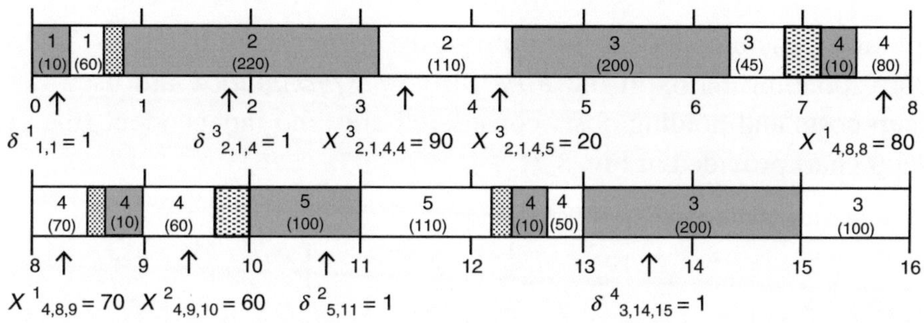

Figure 3.6: Gantt chart for the solution obtained by the proposed model MM1:CLSP-PCSC (for the data given in Sect. 3.5 and Table 3.1) when the setup cost and holding cost of a product are time independent; $Z = 220$ mu

From the Gantt chart it can be observed that Belo-Filho et al.'s model cannot address real-life situations present in process industries such as the production required to be done immediately after setup and uninterrupted production carryover (see Sect. 3.1 for related discussions). Their model doesn't necessarily enforce the starting of production of product i to follow immediately after its setup; in addition, it allows an interruption in the course of production of product i. However, a feasible solution for the *Second proposed formulation* of Belo-Filho et al. could be obtained, when implemented by us (*Note:* Both their models do not address real-life situations present in most process industries such as production starting immediately after setup and uninterrupted production carryover across periods).

When the model developed by Belo-Filho et al. (2013) is executed for time-dependent cost structure, their model (*First proposed formulation*) can be modified for the case where the setup cost is calculated for a product with respect to the period of its setup initiation by modifying their objective function as shown in Eq. (3.61) (see the paper by Belo-Filho et al. (2013) for the variable definitions. Appropriate modifications are done to suit the scenario where setup cost and holding cost of all products are time dependent). However, their model could not be modified for the case where the setup cost is calculated with respect to the period of setup completion.

$$\text{Min } Z = \sum_{i,t,t'<t} bc_i'\,(t'-t)\,d_{i,t'}\,X_{i,t,t'} + \sum_{i,t,t'>t} hc_{i,t}'\,(t'-t)\,d_{i,t'}\,X_{i,t,t'} + \sum_{i,t} sc_{i,t}'\,Z_{i,t} \tag{3.61}$$

For the limited purpose of comparing the model MM1:CLSP-PCSC with the model by Belo-Filho et al. (2013), and also to explain the inadequacy of their model, the same example is used by assuming the setup costs, holding costs and backorders to be time-independent (i.e. in the proposed model, SC_i' is used instead of $SC_{i,t}$ and h_i' is used instead of $h_{i,t}$). When the mathematical model proposed for CLSP-PCSC is executed with these modifications with respect to the objective function, it results with a value of objective function equal to 220 mu. The corresponding Gantt chart is provided in Fig. 3.6. Here, product 4 is produced with two setups (see periods 8, 9 and 10) with a break in production (maintaining the feasibility condition with respect to production occurring immediately after setup and no break in production carryover), and production of all products takes place immediately after their corresponding setup.

Table 3.3: Solution generated by the *First proposed formulation* by Belo-Filho et al. (2013) (corresponding terms in the *First proposed formulation* are used here) when the setup costs and holding costs considered are time independent (the corresponding Gantt chart provided in Fig. 3.7)

$t=1$	$t=2$	$t=3$	$t=4$	$t=5$	$t=6$	$t=7$	$t=8$
$Z_{1,1}=1$ $Z_{2,1}=1$ $X_{1,1,1}=1$ $E_{2,1}=190$ $K_{2,1}=1$	$E_{2,2}=90$ $\alpha_{2,2}=1$ $K_{2,2}=1$ $N_2=1$	$\alpha_{2,3}=1$ $N_3=1$	$X_{2,4,4}=1$ $\alpha_{2,4}=1$ $N_4=1$	$Z_{3,5}=1$ $X_{2,5,5}=1$ $E_{3,5}=155$ $\alpha_{2,5}=1$ $K_{3,5}=1$	$E_{3,6}=55$ $\alpha_{3,6}=1$ $K_{3,6}=1$ $N_6=1$	$X_{3,7,7}=1$ $\alpha_{3,7}=1$ $N_7=1$	$Z_{4,8}=1$ $X_{4,8,8}=1$ $\alpha_{3,8}=1$
$t=9$	$t=10$	$t=11$	$t=12$	$t=13$	$t=14$	$t=15$	$t=16$
$X_{4,9,9}=1$ $\alpha_{4,9}=1$ $N_9=1$	$Z_{5,10}=1$ $X_{4,10,10}=1$ $E_{5,10}=99$ $\alpha_{4,10}=1$ $K_{5,10}=1$	$\alpha_{5,11}=1$ $N_{11}=1$	$X_{5,12,12}=1$ $\alpha_{5,12}=1$ $N_{12}=1$	$Z_{4,13}=1$ $Z_{3,13}=1$ $X_{4,13,13}=1$ $X_{5,13,13}=1$ $E_{3,13}=199$ $\alpha_{5,13}=1$ $K_{3,13}=1$	$E_{3,14}=99$ $\alpha_{3,14}=1$ $K_{3,14}=1$ $N_{14}=1$	$\alpha_{3,15}=1$ $N_{15}=1$	$X_{3,16,16}=1$ $\alpha_{3,16}=1$

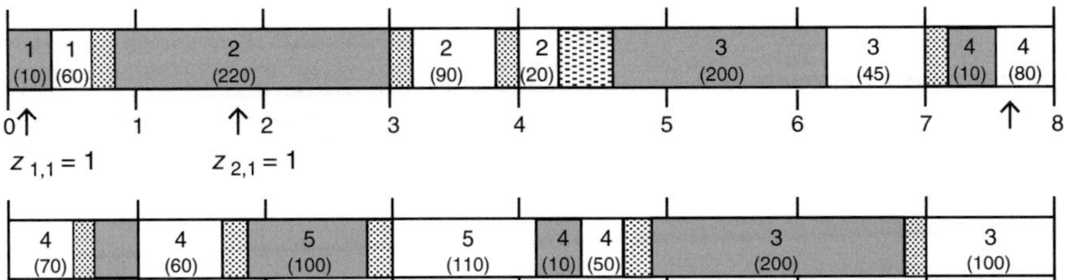

Figure 3.7: Gantt chart for the solution obtained by the *First proposed formulation* by Belo-Filho et al. (2013) (with the proposed modification with respect to one constraint) when the setup costs and holding costs are time independent (for the data given in Table 3.1 and solution given in Table 3.3); $Z = 200$ mu

Remark 2 (with Respect to All Figures): In this book we have illustrated examples with a mix of multiple small setups within a period, and setups extending for more than one period to prove that our models are more generalized in nature. For the sake of illustration the proposed model assumes equal capacity across different periods. However the proposed model allows for different capacities across different time periods due to the use of C_t, without loss of generality.

Here, the value of Z is higher than that of Belo-Filho et al. (2013)'s model because of the two setups which take place to maintain the feasibility with respect to the conditions, production starting immediately after setup and no break in production carryover across periods. *Note:* The model executed, the data used and the result obtained are same as of the special case, 'Setup cost and holding cost of a product are time independent'.

3.6 Proposed Heuristic for CLSP-PCSC with Respect to MM1:CLSP-PCSC

A heuristic is proposed which adapts and possibly improves the idea of the Relax-and-Fix heuristic approach which has been used by authors in the CLSP literature to solve lot sizing problems of large size. The Relax-and-Fix heuristic approach used by some authors is briefly discussed. In the Relax-and-Fix approach used by Belvaux and Wolsey (2000), a time window of a particular duration is assumed (e.g. Time window = 4 time periods). Here, in the first time window (first four periods), all the variables declared as binary variables are restricted to be binary, and all the other binary variables in the rest of the time horizon are relaxed so as to be continuous. With this setting, this particular sub-problem is solved using an MIP solver, and in the next step, by using the solution obtained from the first sub-problem, the binary variables within the first time window are fixed. In the next sub-problem the next time window is considered (periods five to eight) and the approach is repeated till the last time window is completed. In the Relax-and-Fix algorithm used by Stadtler (2003), the time horizon is divided into three parts. The first part consists of a time window of a particular size (e.g. α = four periods); the second part contains periods preceding that time window and the third part contain periods which follow that time window. Within the time window of size α, the first few periods are selected (for example, periods 1, 2 and 3 are selected) and the binary decision variables in those periods are fixed as binary and binary decision variables in remaining part of the time window are relaxed to be continuous (let the size of the remaining periods within the time window be defined as γ). After solving this sub-problem using an MIP solver, the binary variables in first few periods of size (let the size be β) $\beta \leq \alpha$ - γ are fixed with the resulting binary variables (solution) obtained. The next sub-problem to be solved starts from period β + 1 and ends in period β + α. These steps are followed until all the binary variables are fixed in the entire time horizon. In the Relax-and-Fix heuristic proposed by Akartunalı and Miller (2009), the first sub-problem is solved similar to Belvaux and Wolsey (2000). But while solving the second sub-problem, not all the binary variables within the first time window (e.g. five periods) are fixed from the resulting binary variables (solution) of the first sub-problem. Only the variables present in a subset of the time periods (e.g. periods 1, 2 and 3) in the first time window (of the first sub-problem) are fixed using the resulting solution. Also, the time window considered in the second sub-problem (in which all binary variables are restricted to be binary $\in \{0, 1\}$) overlaps with the time window of the first sub-problem. The time window considered in the second sub-problem contains some time periods of the first sub-problem in which the binary variables (solution) are not frozen (e.g. periods 4 and 5) and some periods succeeding the first time window (e.g. periods 6, 7 and 8). The second sub-problem is again solved in a similar way, and the process continues till all the binary variables are fixed in the entire time horizon. In the Relax-and-Fix heuristic developed by Toledo et al. (2015), initially all the variables are relaxed. According to them, a time window of a particular size is selected next, and all binary variables within the window are fixed as binary integers, while the other binary variables outside the time window are relaxed. After solving this sub-problem, a subset of the binary variables from the resulting solution are fixed, and another set of integer and relaxed variables are optimized. Three types of time windows are proposed: the row-wise time window, in which the window moves along rows; column-wise time window, in which the window moves along columns and value-wise time window, in which the window consists of variables with relaxed values closest to the value 0.5.

The proposed heuristic in this work uses the concept of Relax-and-Fix with limited similarity to the related works of the authors mentioned before; however, the proposed heuristic has its own novelty especially in terms of determining the size of the time window and solving each sub-problem. The proposed CLSP-PCSC heuristic is illustrated with the help of Fig. 3.8 and the steps are briefly explained using a pseudocode. Here $MoveWindow$ refers to the window up to which all the binary variables are set to be binary integers, starting from period 1, i.e., all the binary variables $\in \{0, 1\}$ (B_{MIP} refers to binary integers set as rigid binary variables in $MoveWindow$), and $FinWindow$ is the window in which all the binary integer variables are relaxed as continuous variables within the interval [0,1], up to period T (LP_R refers to relaxation of binary variables in all periods after $MoveWindow$). $WinSize$ refers to the size of the window (expressed in time periods) which is assigned to $MoveWindow$. The size of a time window for every problem instance (see Eq. (3.62)) is expressed as the ceiling value or the smallest integer value greater than or equal to the ⌈maximum setup time (among all products expressed in time units) divided by the minimum capacity (among all periods expressed in time units)⌉ (rounded off to the higher integer if necessary)+1 (expressed in time periods).

$$WinSize \text{ (expressed in time periods)} = \left\lceil \frac{max_i\{ST_i\}\text{(expressed in time units)}}{min_t\{C_t\}\text{(expressed in time units)}} \right\rceil + 1 \quad (3.62)$$

The symbol τ (initially set to zero) is used to refer to the time period up to which the binary variables are set before solving a sub-problem. Once a sub-problem is solved, the binary (solution) variables (equal to one) which appear in the resulting solution of the sub-problem are considered, i.e., while solving the subsequent sub-problem. In the subsequent sub-problem, the set of inequalities $SolnEqu_{1,i,t,\tau}$, $SolnEqu_{2,i,t,\tau}$, $SolnEqu_{3,i,t,\tau}$ and $SolnEqu_{4,i,t,\tau}$ are introduced appropriately with respect to each binary variable (solution) that is equal to one appearing in the preceding solution (until period τ). These respective inequalities are set (i.e. invoked), while solving every sub-problem by using the mathematical model. These aspects are the highlights of the proposed heuristic.

First Sub-Problem: While solving the first sub-problem, τ takes the value zero, which means that none of the periods are preset with any inequality ($SolnEqu's$) in the first iteration. In the first sub-problem, all binary variables (in $MoveWindow$) $\in \{0,1\}$ and all the binary variables in $FinWindow$ are relaxed in the interval [0,1], and the model is solved using an MIP solver.

Subsequent Sub-Problem Until $(T - WinSize + 1)$*th Sub-Problem:* In the subsequent sub-problems, τ is incremented to τ+1. Using the resulting solution values from the preceding sub-problem (i.e. the binary variables with the value 1 present up to the τth time period obtained from the previous sub-problem) are considered, and each such binary variable leads to the corresponding introduction to one of $SolnEqu_{1,i,t,\tau}$, $SolnEqu_{2,i,t,\tau}$, $SolnEqu_{3,i,t,\tau}$ and $SolnEqu_{4,i,t,\tau}$, as appropriate, in the mathematical model; such binary variables are now set as $\in \{0,1\}$ and not equal to 1, while solving the present sub-problem. The corresponding inequalities are set (i.e. invoked) while solving the current sub-problem due to which, either a left shift or a right shift or *status quo* of a setup/production is allowed to occur, while solving the subsequent sub-problems, so as to possibly improve the solution. Also, such inequalities are introduced to ensure at least one setup initiation of the corresponding product i appropriately in periods $t-1$, t or $t+1$. Therefore, while solving the current sub-problem, periods from $\tau+1$ to $\tau + WinSize$ are considered as the next $MoveWindow$ up to which all the binary variables are considered as belonging to $\{0,1\}$ right from time period 1; and $FinWindow$ starts from period $\tau + WinSize + 1$ and goes up to period T where all binary variables are relaxed as continuous variables in the interval [0,1]. Note that $SolnEqu_{1,i,t,\tau}$, $SolnEqu_{2,i,t,\tau}$, $SolnEqu_{3,i,t,\tau}$ and $SolnEqu_{4,i,t,\tau}$ are not considered in the $MoveWindow$. Such a sub-problem (with such a setting of inequalities up to period τ, setting of binary variables up to $\tau + WinSize \in \{0,1\}$ from period 1, and all binary variables relaxed as continuous variables in the interval [0,1] from period $(\tau + WinSize + 1)$ up to period T) is solved using an MIP solver. This procedure continues until all sub-problems are solved up to the complete time horizon. It should be noted that binary variables are present as either 0 or 1 (without being relaxed) in the final solution. The number of sub-problems to be solved for every instance in the proposed heuristic is equal to $T - WinSize + 1$.

It is evident that the proposed heuristic employs a set of inequalities corresponding to the binary values (that take the value equal to one) obtained up to the previous sub-problem, instead of using the binary variables (with the rigid setting of 1) directly. This aspect is unique to this work. The purpose of using inequalities instead of using the strict binary values is that the use of inequalities improves the quality of the solution obtained, by allowing for a possible right shift or left shift or *status quo* with respect to the corresponding setup initiation of a product.

The principle behind using inequalities instead of involving binary variables directly is explained using an example. Consider the example as follows: the problem consists of determining the production quantities for a five product, six period problem; the setup costs (mu/setup) of the products are assumed to be time independent, given by $SC_1 = 50$, $SC_2 = 10$, $SC_3 = 30$, $SC_4 = 20$ and $SC_5 = 40$; the holding costs (mu/period/unit product carried over) are $h_1 = 2$, $h_2 = 3$, $h_3 = 1$, $h_4 = 1$ and $h_5 = 2$; the backorder costs (mu/period/unit product backordered) for all products are assumed to have a high value equal to 50,000; the capacity (time units) for all periods is $C_t = 100\ \forall t$; $a_i = 1\ \forall i$ (time unit/unit of product i) corresponds to the number of time units required for producing one unit of product i; the setup time (time units) of the products is assumed to be $ST_1 = 10$, $ST_2 = 110$, $ST_3 = 10$, $ST_4 = 20$ and $ST_5 = 100$; and the demand (units) for all products is $d_{1,1} = 60$, $d_{2,3} = 40$, $d_{3,3} = 40$, $d_{4,4} = 70$, $d_{1,6} = 40$ and $d_{5,6} = 100$. Note: Equal capacity is assumed across all periods for the sake of illustration. Without loss of generality, the proposed model and heuristic allows different capacities across different periods. Also, in this problem Eq. (3.62) of the proposed mathematical model is included, and it is assumed that the demand for all products should be satisfied within period T.

When the problem is executed using the proposed exact method (i.e. executing model MM1: CLSP-PCSC) the value of the objective function obtained is Z = 500,530 mu. It can be noted in this numerical illustration that backorder of a product is inevitable in view of high backorder cost, due to which the value of Z = 500,530 mu.

The proposed heuristic uses inequalities in each sub-problem, instead of setting or "freezing" the binary variables as 1, wherever present up to a given sub-problem. A comparison involving the use of binary variables (solution) obtained by freezing those values versus the introduction of the $SolnEqu's$ inequalities corresponding to the binary variables (solution) obtained, while solving a sub-problem in the proposed heuristic, is illustrated in Fig. 3.9. In the problem instance considered, when a sub-problem is solved using the proposed heuristic (however, by "freezing" the values of the binary variables (solution) obtained up to period τ, in the previous sub-problem), an integer infeasible solution is obtained after solving the last sub-problem (see Fig. 3.9a). This is so because "freezing" binary variables (solution) while solving a particular sub-problem does not allow the heuristic to search for feasible solutions (or improved feasible solutions) by allowing for the shifting of setups to the left or the right periods, if a better solution can be obtained. When the problem is solved using the proposed heuristic method (including the inequalities that correspond to the binary variables with value equal to 1, present in the pseudocode until period τ), the value of the objective function obtained is Z = 500,530 mu, associated with feasibility. The proposed heuristic, using the inequalities corresponding to the binary values (solution) obtained (i.e. binary variables taking the value of 1), is explained step by step in Fig. 3.9b.

$MoveWindow$ is set to the value of $WinSize$ of size 3 time periods, and all the binary variables in this $MoveWindow$ are set as $\in \{0, 1\}$, and in $FinWindow$ that starts from period $\tau + WinSize + 1$ up to period T, all binary variables are relaxed (time periods 4–6). The first sub-problem (i.e. when $\tau = 0$) is solved using the above conditions. Once this sub-problem is solved, $\delta^1_{1,1} = 1$ is generated in the resulting solution. In the second sub-problem (with $\tau = 1$), $SolnEqu_{1,i,t,\tau}$ is introduced appropriately with respect to the value of the corresponding binary variable (solution) $\delta^1_{1,1} = 1$ (obtained in the first period ($t = 1$) of the first sub-problem (where $i = 1$)) into the current sub-problem and then this sub-problem is solved.

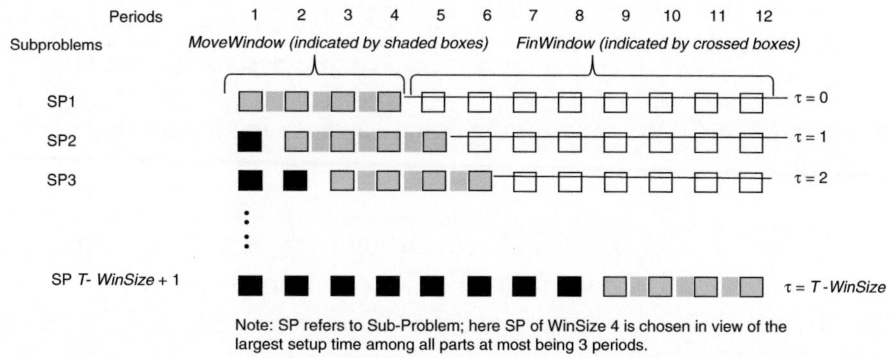

Figure 3.8: Figure illustrating the CLSP-PCSC heuristic

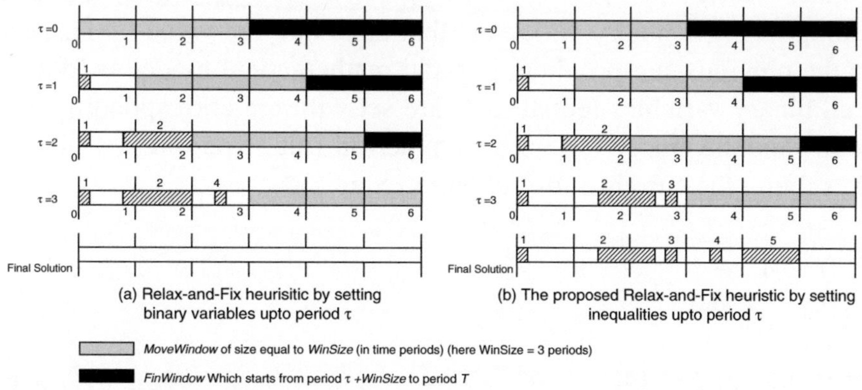

Figure 3.9: Figure illustrating an example (also discussed in the text) where binary variables are set up to period τ ((**a**) and inequalities are set up to period τ (**b**)), in the heuristic proposed based on MM1:CLSP-PCSC

Remark 3 (with Respect to Fig. 3.9): All binary variables up to ($\tau + WinSize$) are set as $\in \{0, 1\}$ and all binary variables from $\tau + WinSize + 1$ to T are set to [0,1], i.e., relaxed in the interval 0 to 1.

Pseudocode for the Proposed Heuristic Based on MM1:CLSP-PCSC

1: **function**

Heuristic ($MoveWindow, FinWindow, WinSize, \tau, t, C', SolnEqu_{1,i,t,\tau}, SolnEqu_{2,i,t,\tau}, SolnEqu_{3,i,t,\tau}, SolnEqu_{4,i,t,\tau}$)

2: **Compute**

$$WinSize(\text{expressed in time units}) = \left\lceil \frac{max_i\{ST_i\}(\text{expressed in time units})}{C'(\text{expressed in time units})} \right\rceil + 1$$

Note: Here, $C' = C_t \;\; \forall t$ (assuming equal capacity in all periods)

3: **while** ($\tau = 0$) **do**

{

4: $MoveWindow \longleftarrow$ of size $WinSize$ starting from period 1 to period $\tau + WinSize$

5: $FinWindow \longleftarrow$ all periods from $WinSize + 1$ up to period T

6: **Solve** $MoveWindow \longleftarrow B_{MIP}$ and $FinWindow \longleftarrow LP_R$

} /* End of while loop */

7: $\tau = 1$

8: **while** ($\tau \leq T - WinSize$) **do**

{

9: Introduce $SolnEqu_{1,i,t,\tau}, SolnEqu_{2,i,t,\tau}, SolnEqu_{3,i,t,\tau}$ and $SolnEqu_{4,i,t,\tau}$ corresponding to those binary variables (that have had the value of 1), generated up to τ from the previous sub-problem, into the mathematical model as follows. /*It implies that all binary variables (equal to 1) are set with the corresponding $SolnEqu$'s up to period τ, and the binary variables from period 1 up to period $(\tau + WinSize) \in \{0, 1\}$, while solving the current sub-problem in Step 12*/:

9.1: $SolnEqu_{1,i,t,\tau} \longleftarrow (\delta^1_{i,t-1} + \delta^2_{i,t-1} + \delta^3_{i,t-1,t} + \delta^1_{i,t} + \delta^2_{i,t} + \delta^3_{i,t,t+1} \geq 1)$
$\forall i \mid \delta^1_{i,t} = 1$ and $\forall t \leq \tau$;

9.2: $SolnEqu_{2,i,t,\tau} \longleftarrow (\delta^2_{i,t-1} + \delta^3_{i,t-1,t} + \delta^1_{i,t} + \delta^2_{i,t} + \delta^3_{i,t,t+1} + \delta^1_{i,t+1} + \delta^2_{i,t+1} + \delta^3_{i,t+1,t+2} \geq 1)$
$\forall i \mid \delta^2_{i,t} = 1$ and $\forall t \leq \tau$;

9.3: $SolnEqu_{3,i,t,\tau} \longleftarrow (\delta^2_{i,t-1} + \delta^3_{i,t-1,t} + \delta^1_{i,t} + \delta^2_{i,t} + \delta^3_{i,t,t+1} + \delta^1_{i,t+1} + \delta^2_{i,t+1} + \delta^3_{i,t+1,t+2} \geq 1)$
$\forall i \mid (\delta^3_{i,t,t+1} = 1; \sum_{t'=t+2}^{T} \delta^3_{i,t,t'} = 0$ and $ST_i \leq C')$ and $\forall t \leq \tau$;

9.4: $SolnEqu_{3,i,t,\tau} \longleftarrow (\delta^3_{i,t-1,t'-1} + \delta^4_{i,t-1,t'-1} + \delta^3_{i,t,t'} + \delta^4_{i,t,t'} + \delta^3_{i,t+1,t'+1} + \delta^4_{i,t+1,t'+1} \geq 1)$
$\forall i \mid (\sum_{t'=t+1}^{T} \delta^3_{i,t,t'} = 1$ and $ST_i \geq C')$ and $\forall t \leq \tau$;

9.5: $SolnEqu_{4,i,t,\tau} \longleftarrow (\delta^3_{i,t-1,t'-1} + \delta^4_{i,t-1,t'-1} + \delta^3_{i,t,t'} + \delta^4_{i,t,t'} + \delta^3_{i,t+1,t'+1} + \delta^4_{i,t+1,t'+1} \geq 1)$
$\forall i \mid (\sum_{t'=t+1}^{T-1} \delta^4_{i,t,t'} = 1$ and $ST_i \geq C')$ and $\forall t \leq \tau$.

Continued... Pseudocode for the Proposed Heuristic Based on MM1:CLSP-PCSC

10: $MoveWindow \longleftarrow$ all periods starting from period $\tau + 1$ to period $\tau + WinSize$
11: $FinWindow \longleftarrow$ all periods from $\tau + WinSize + 1$ up to period T
12: **Solve** $MoveWindow \longleftarrow B_{MIP}$, $FinWindow \longleftarrow LP_R$ and with $SolnEqu_{1,i,t,\tau}$, $SolnEqu_{2,i,t,\tau}$, $SolnEqu_{3,i,t,\tau}$ and $SolnEqu_{4,i,t,\tau}$ as appropriate.

Note: $FinWindow$ exists up to $\tau = T - WinSize - 1$, and not when $\tau = T - WinSize$. Therefore, when $\tau = T - WinSize$ the heuristic is solved without including $FinWindow \longleftarrow LP_R$. Also, $MoveWindow$ does not include the consideration of $SolnEqu_{1,i,t,\tau}$, $SolnEqu_{2,i,t,\tau}$, $SolnEqu_{3,i,t,\tau}$ and $SolnEqu_{4,i,t,\tau}$.

13: $\tau = \tau + 1$ } /* End of while loop */
14: **end function**

Note: $\delta^1_{i,0} = \delta^2_{i,0} = 0\ \forall i$; $\delta^3_{i,0,t} = \delta^4_{i,0,t} = 0\ \forall i$ and $\forall t$, and $\delta^4_{i,T-1,T} = 0\ \ \forall i$ (boundary conditions).

Therefore, $SolnEqu_{1,1,1,1} \longleftarrow (\delta^1_{1,0} + \delta^2_{1,0} + \delta^3_{1,0,1} + \delta^1_{1,1} + \delta^2_{1,1} + \delta^3_{1,1,2} \geq 1)$ where $\delta^1_{1,0} = \delta^2_{1,0} = \delta^3_{1,0,1} = 0$ as boundary conditions.

Once the second sub-problem is solved, the resulting solution has $\delta^1_{1,1} = 1$ and $\delta^3_{2,1,2} = 1$ in the first time period and no binary variables are present in the second time period. In the third sub-problem (with $\tau = 2$), $SolnEqu_{1,i,t,\tau}$ (where $i = 1$ and $t = 1$) and $SolnEqu_{4,i,t,\tau}$ (where $i = 2$ and $t = 1$) are introduced appropriately with respect to the value of the corresponding binary variables (solution) $\delta^1_{1,1} = 1$ and $\delta^4_{2,1,2} = 1$ (generated in the resulting solution of the second sub-problem) into the current sub-problem and then this sub-problem is solved.

$SolnEqu_{1,1,1,2} \longleftarrow (\delta^1_{1,0} + \delta^2_{1,0} + \delta^3_{1,0,1} + \delta^1_{1,1} + \delta^2_{1,1} + \delta^3_{1,1,2} \geq 1)$ where $\delta^1_{1,0} = \delta^2_{1,0} = \delta^3_{1,0,1} = 0$, and

$SolnEqu_{4,2,1,2} \longleftarrow (\delta^3_{2,0,1} + \delta^4_{2,0,1} + \delta^3_{2,1,2} + \delta^4_{2,1,2} + \delta^3_{2,2,3} + \delta^4_{2,2,3} \geq 1)$ where $\delta^3_{2,0,1} = \delta^4_{2,0,1} = 0$.

Once the third sub-problem is solved, the resulting solution has $\delta^1_{1,1} = 1$ in the first time period, $\delta^3_{2,2,3} = 1$ in the second time period and $\delta^1_{3,3} = 1$ in the third time period. In the fourth sub-problem (with $\tau = 3$), $SolnEqu_{1,i,t,\tau}$ where $i = 1$, $t = 1$ and $\tau = 3$; $SolnEqu_{3,i,t,\tau}$ where $i = 2$, $t = 2$ and $\tau = 3$ and $SolnEqu_{1,i,t,\tau}$ where $i = 3$, $t = 3$ and $\tau = 3$ are introduced appropriately with respect to the value of the corresponding binary variables (solution) $\delta^1_{1,1} = 1$, $\delta^3_{2,2,3} = 1$ and $\delta^1_{3,3} = 1$ (generated in the first three periods of the third sub-problem) into the current sub-problem and then this sub-problem is solved (see below for the settings corresponding to the binary variables).

$SolnEqu_{1,1,1,3} \longleftarrow (\delta^1_{1,0} + \delta^2_{1,0} + \delta^3_{1,0,1} + \delta^1_{1,1} + \delta^2_{1,1} + \delta^3_{1,1,2} \geq 1)$ where $\delta^1_{1,0} = \delta^2_{1,0} = \delta^3_{1,0,1} = 0$,

$SolnEqu_{3,2,2,3} \longleftarrow (\delta^3_{2,1,2} + \delta^4_{2,1,2} + \delta^3_{2,2,3} + \delta^4_{2,2,3} + \delta^3_{2,3,4} + \delta^4_{2,3,4} \geq 1)$, and

$SolnEqu_{3,1,3,3} \longleftarrow (\delta^1_{3,2} + \delta^2_{3,2} + \delta^3_{3,2,3} + \delta^1_{3,3} + \delta^2_{3,3} + \delta^3_{3,3,4} \geq 1)$.

The resulting solution of the fourth sub-problem consists of binary variables $\delta^1_{1,1} = 1$, $\delta^3_{2,2,3} = 1$, $\delta^1_{3,3} = 1$, $\delta^1_{4,4} = 1$ and $\delta^2_{5,5} = 1$.

Note that the heuristic presented for MM1:CLSP-PCSC assumes identical C_t across all time periods, where $C_t = 100$ (time units), and the possible shift of setup/production is restricted to one period ahead of or after the current time period or *status quo*. However, when non-identical C_t's is present across time periods, allowing the possible shift of setup/production for more periods ahead of or after the current time period, the extension of the heuristic based on MM1:CLSP-PCSC becomes tedious. In such cases, with these considerations, a second mathematical model and a comprehensive heuristics are developed. The details are given in Chap. 4.

3.7 Computational Experience

In this section, the solution (CPU) times of the proposed MM1:CLSP-PCSC model are given. Following this, the proposed mathematical model MM1:CLSP-PCSC is compared with the heuristic proposed with respect to MM1:CLSP-PCSC.

3.7.1 Comparing Solution Times of the Proposed Mathematical Models (MM1:CLSP-PCSC)

In Table 3.4, the solution time (to obtain an optimal solution) obtained by executing the generalized MM1:CLSP-PCSC for each of the 162 sample problem instances of various product and time-period combinations is presented. All the problem instances have been executed on a Pentium 3.10 GHz Windows 7 workstation with 4.00 GB RAM, using CPLEX v12.4. It is believed that if the same problem instances are run on the latest version of CPLEX (v12.6.3), the solution times will improve. While using CPLEX, the values of the mixed integer optimality gap tolerance and integrality tolerance were set to zero, in order to obtain accuracy in results and maintain the integrality with respect to binary variables. Alternate optimal solutions are obtained for the problem instances when the default tolerance values of the CPLEX are retained.

The two tolerances which had been set to zero instead of using the default values are the integrality tolerance and the MIP tolerance. The parameter 'integrality tolerance' sets the amount by which a computed solution value for an integer variable can violate integrality; it does not specify an amount by which CPLEX relaxes integrality. In order to avoid violation of integrality and to obtain the exact values without floating the integer solution values, it has been set it exactly to zero such that the resulting solution values are exact and rounding off the solution values need not be done manually. Sometimes having tolerance set to zero helps to compare our results with other exact methods (existing mathematical models addressing the same problem), so that no variables (which need not be necessarily float) are present in the final solution. Therefore, in order to obtain the exact values of the variables, the tolerance has been set to value 0 while executing the models using the solver. The parameter MIP tolerance is the gap between the best integer objective and the objective of the best node remaining. When this difference falls below the value of this parameter, the mixed integer optimization is stopped. In order to stop the optimization only at the best integer objective, the tolerance limit has been set to zero while executing the models using the solver.

The following settings are assumed while performing the computations: The time-dependent setup costs (mu/setup) of the products range from 5 mu/setup to 60 mu/setup; the time-dependent holding costs (mu/period/unit product carried over) of the products range from 1 mu/period/unit product carried over to 5 mu/period/unit product carried over; the setup time (time units) of the products (per setup) ranges from 5 time units to 460 time units and the backorder cost (mu/period/unit product backordered) of all products is assumed to be 5,000,000 mu/period/unit product backordered (which implies no backorder). Throughout this work 100 time units correspond to one time period (t) and the number of time units required for producing one unit of product i (a_i) is equal to 1 time unit per unit of product i.

The problems of various sizes which were executed with the above range of settings seem computationally fast. The computational effort does depend upon the capacity available versus demand of the products across the planning horizon. When the initial computational investigation was carried out it was found out that when there is less demand and the capacity is fairly large (and hence resulting in the idle time of the machine resource), there is no need of setup crossover. Also, when there is more demand and less capacity, we have infeasibility with respect to setup crossover. Among the various problem instance generated, Table 3.4 lists only the solution times of the optimal problem instances of various sizes. Note: In most of the problem instances executed, when the capacity was fairly large compared to the demand, the CPU time was much less in comparison to the situation where the capacity availability and demand requirement were fairly tight.

3.7.2 Comparison of Exact and Heuristic Approaches of MM1:CLSP-PCSC

The computational experience for the proposed heuristic is given in Table 3.5, and it is compared with the results of the proposed exact method (MM1:CLSP-PCSC). It can be observed that the associated computational effort is quite good. The instances used in the computational evaluation are given in Tables 3.6, 3.7, 3.8, 3.9 along with their Gantt charts given in Figs. 3.10, 3.11, 3.12, 3.13, 3.14, 3.15, 3.16, 3.17, 3.18, 3.19, 3.20, 3.21, 3.22, 3.23, 3.24, 3.25. Equal capacity is assumed in all periods in these problem instances, where $C_t = 100$ (time units). Equal capacity is assumed across all periods for the sake of numerical illustration. However, the proposed heuristic approach allows different capacities across different periods, without loss of generality. MM2:CLSP-PCSC also results in the same value of Z for the mathematical model and the heuristic approach. "Hard" problems have been considered in the proposed heuristic and it is hence ensured that the capacity availability and demand requirement were almost matching in every period (as we have tight constraints) in order to get the fair information about the computational performance of the proposed heuristic.

Table 3.4: Computational time (in sec.) for various problem instances

Description	Number of Periods / Number of Products	4	6	8	10	12	14	16	18	20
Instance 1	4	0.03	0.08	0.09	0.19	2.71	0.86	4.96	5.65	11.84
Instance 2		0.03	0.06	0.45	0.64	0.76	0.61	1.78	4.09	3.82
Average		0.03	0.07	0.27	0.42	1.74	0.74	3.37	4.87	7.85
Instance 1	6	0.05	0.09	0.59	0.53	0.72	1.08	1.79	10.17	9.08
Instance 2		0.03	0.06	0.22	0.30	0.61	1.14	2.59	2.89	39.78
Average		0.04	0.08	0.41	0.42	0.67	1.11	2.19	6.53	24.43
Instance 1	8	0.06	0.13	0.22	0.42	2.73	1.68	2.51	30.87	8.92
Instance 2		0.05	0.13	0.38	0.45	2.53	1.70	2.64	4.76	6.72
Average		0.06	0.13	0.30	0.44	2.63	1.69	2.58	17.82	7.82
Instance 1	10	0.05	0.45	0.31	0.72	1.30	2.42	7.53	6.80	12.84
Instance 2		0.05	0.17	0.50	3.34	2.14	2.59	3.93	6.04	9.88
Average		0.05	0.31	0.41	2.03	1.59	2.51	5.73	6.42	11.36
Instance 1	12	0.09	0.25	0.53	0.78	2.34	3.32	15.38	50.17	18.17
Instance 2		0.20	0.17	0.51	1.75	1.87	3.53	28.27	11.00	20.64
Average		0.15	0.21	0.52	1.27	2.11	3.43	21.83	30.59	19.41
Instance 1	14	0.17	0.39	0.97	1.31	14.09	5.82	7.50	14.31	18.53
Instance 2		0.26	0.23	0.72	4.60	3.48	4.46	13.38	16.68	445.77
Average		0.22	0.31	0.85	2.96	8.79	5.14	10.44	15.50	232.15
Instance 1	16	0.11	0.69	1.45	1.39	3.38	5.60	10.02	20.36	497.80
Instance 2		0.19	0.28	1.01	4.26	1.14	5.60	10.92	36.24	29.60
Average		0.15	0.49	1.23	2.83	2.26	5.60	10.47	28.30	263.70
Instance 1	18	0.69	0.33	1.20	3.03	5.46	19.84	13.51	19.72	56.86
Instance 2		0.13	0.67	1.00	4.38	9.45	7.21	16.26	18.75	94.02
Average		0.41	0.5	1.1	3.71	7.46	13.53	14.89	19.24	75.44
Instance 1	20	0.61	0.47	1.36	2.82	4.74	13.17	38.61	31.82	102.20
Instance 2		0.27	0.44	2.21	10.33	5.32	8.64	13.00	19.45	96.03
Average		0.44	0.46	1.79	6.58	5.03	10.62	25.81	25.64	99.12

Table 3.5: Computational experience for the exact and heuristic approaches of MM1:CLSP-PCSC

Backorder cost	Methodology	Description	Products			
			10	12	14	16
1st set	Exact	Z (mu)	311010.99	1062095.99	632345.00	1381185.00
		Solution time (sec)	493.82	9989.76	20580.10	80365.30
	Heuristic	Z (mu)	340775.50	1081985.99	681275.00	2060665.00
		Solution time (sec)	238.06	425.36	771.25	1172.32
2nd set	Exact	Z (mu)	4110.00	12490.12	8075.00	14985.00
		Solution time (sec)	303.98	14192.25	2109.71	57721.99
	Heuristic	Z (mu)	4170.12	12625.10	8075.00	21265.00
		Solution time (sec)	243.58	455.94	729.33	964.04

Note:
1st set—refers to the set of backorder costs with values in thousands (monetary units), with T=20 time periods
2nd set—refers to the set of backorder costs with values in tens (monetary units), with T=20 time periods
Also note: In a cell, in the following Tables 3.6, 3.7, 3.8 and 3.9;

$$\begin{cases} a & \text{a refers to demand } (d_{i,t}); \\ b & \text{b refers to data pertaining to time-dependent setup cost } (SC_{i,t}) \text{ (mu/setup)}; \\ c & \text{c refers to data pertaining to time-dependent holding cost } (h_{i,t}) \text{ (mu/period/unit product carried over)}. \end{cases}$$

Table 3.6: Data for ten products and twenty time periods, with two different sets of backorders costs. The Gantt charts corresponding to the exact and heuristic solutions considering the 1st set of backorder costs are given in Figs. 3.10 and 3.11; and the Gantt charts corresponding to the 2nd set of backorder costs are given in Figs. 3.12 and 3.13

Product	Setup time (ST_i): (time units)	1st set of backorder costs (b_i): (mu/period/ unit product carried over)	2nd set of backorder costs (b_i): (mu/period/ unit product carried over)	Number of time units required for producing one unit of product i (a_i): (time units/ unit of product i)
1	10	2000	20	1
2	10	3000	30	1
3	20	1000	10	1
4	100	1000	10	1
5	120	2000	20	1
6	20	4000	40	1
7	10	3000	30	1
8	10	1000	10	1
9	110	2000	20	1
10	220	4000	40	1

Demand ($d_{i,t}$), Setup cost ($SC_{i,t}$): (mu/setup) and Holding cost ($h_{i,t}$): (mu/period/unit product carried over)

Product \ Period	1	2	3	4	5	6	7	8	9	10	11	12	13	14	15	16	17	18	19	20
1	20	0	0	0	0	0	0	0	0	0	0	0	30	0	40	0	0	0	0	40
	20	30	10	30	50	20	40	20	40	30	20	10	5	60	30	20	30	10	25	40
	1	2	3	1	2	1	2	1	3	2	1	3	2	1	2	1	1	3	2	1
2	40	0	0	0	0	0	0	0	0	0	0	0	0	20	40	0	0	0	0	0
	40	20	40	30	20	10	5	60	30	20	20	30	10	30	50	60	60	30	20	20
	1	3	2	1	3	2	1	2	1	1	2	3	1	2	1	2	2	3	1	2
3	0	90	0	0	0	0	0	0	0	0	0	0	0	0	0	0	0	0	0	40
	30	50	60	40	20	40	30	20	20	30	15	10	5	60	30	20	20	40	30	20
	1	2	1	3	2	1	3	2	1	2	1	1	2	3	1	2	1	1	2	3

(continued)

Table 3.6: (continued)

Product \ Period	1	2	3	4	5	6	7	8	9	10	11	12	13	14	15	16	17	18	19	20
	Demand ($d_{i,t}$), Setup cost ($SC_{i,t}$): (mu/setup) and Holding cost ($h_{i,t}$): (mu/period/unit product carried over)																			
4	0	40	50	0	0	0	0	0	0	0	0	0	0	0	0	0	0	0	0	0
	20	40	30	20	10	5	60	30	20	20	30	10	30	50	60	40	30	50	60	40
	1	2	1	2	1	3	2	1	2	3	1	3	2	1	2	1	1	3	2	1
5	0	0	0	0	0	80	0	0	0	0	0	0	0	0	0	0	0	0	0	0
	60	30	20	20	30	15	30	50	60	40	20	40	30	20	10	5	40	20	40	30
	1	2	1	1	3	2	1	2	3	1	2	1	2	1	3	2	2	1	2	1
6	0	0	0	0	0	0	70	0	0	0	0	0	0	0	0	0	0	0	0	0
	30	10	25	40	50	45	60	70	15	20	35	55	10	30	50	60	20	30	10	30
	2	1	1	2	3	1	2	1	2	2	3	1	2	1	2	1	2	1	3	2
7	0	0	0	0	0	0	0	20	0	0	0	0	0	50	0	90	0	0	0	0
	20	40	30	20	20	30	15	10	5	60	30	20	20	40	20	30	10	30	50	20
	1	3	2	1	2	1	1	2	3	1	2	1	1	2	3	2	2	3	1	2
8	0	0	0	0	0	0	0	50	0	0	0	0	0	0	0	0	0	0	0	0
	20	20	30	15	30	50	60	40	20	40	30	20	10	5	40	20	40	20	30	15
	1	2	1	2	1	3	2	1	2	3	1	3	2	2	1	3	2	2	1	2
9	0	0	0	0	0	0	0	10	50	40	0	0	0	0	0	0	0	0	0	0
	20	30	15	30	50	60	40	20	40	30	20	10	5	40	50	20	40	20	40	30
	2	1	3	2	1	2	1	2	1	2	2	1	2	1	2	1	3	2	2	1
10	0	0	0	0	0	0	0	0	0	0	0	0	40	0	0	10	30	30	10	0
	20	40	30	20	10	5	60	30	20	20	30	10	20	20	30	15	30	50	60	40
	3	1	1	2	3	2	3	2	2	1	2	3	1	2	1	3	2	1	3	2
C_t	100	100	100	100	100	100	100	100	100	100	100	100	100	100	100	100	100	100	100	100

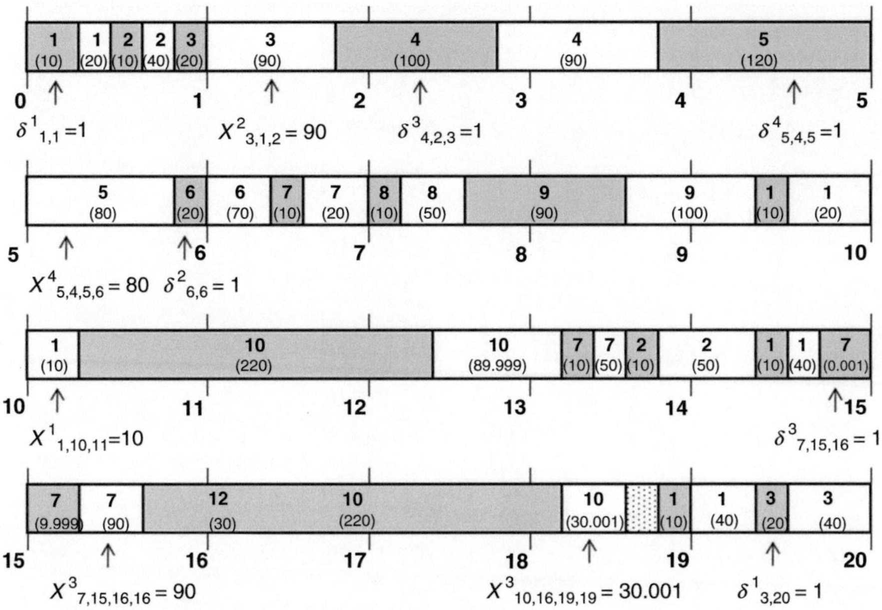

Figure 3.10: Gantt chart for the solution obtained by the proposed model MM1:CLSP-PCSC considering ten products and twenty time periods, when the setup costs and holding costs are product dependent and time dependent, with the setup cost for a product calculated with respect to the period of its setup initiation (for the data given in Table 3.6) considering the 1st set of backorder costs; $Z = 311010.99$ mu

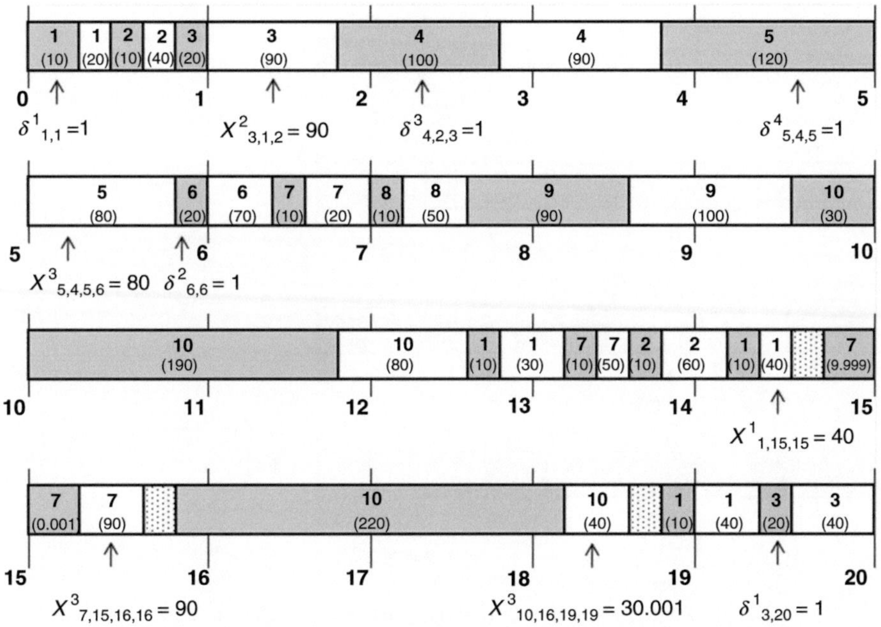

Figure 3.11: Gantt chart for the solution obtained by the proposed heuristic considering ten products and twenty time periods, when the setup costs and holding costs are product dependent and time dependent, with the setup cost for a product calculated with respect to the period of its setup initiation (for the data given in Table 3.6) considering the 1stset of backorder costs; $Z = 340777.50$ mu

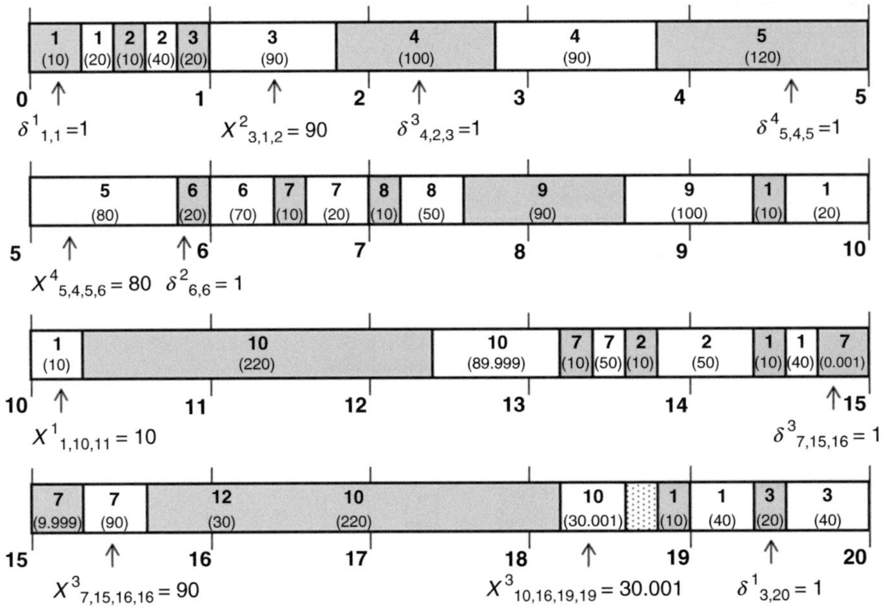

Figure 3.12: Gantt chart for the solution obtained by the proposed model MM1:CLSP-PCSC considering ten products and twenty time periods, when the setup costs and holding costs are product dependent and time dependent, with the setup cost for a product calculated with respect to the period of its setup initiation (for the data given in Table 3.6) considering the 2nd set of backorder costs; $Z = 4110.00$ mu

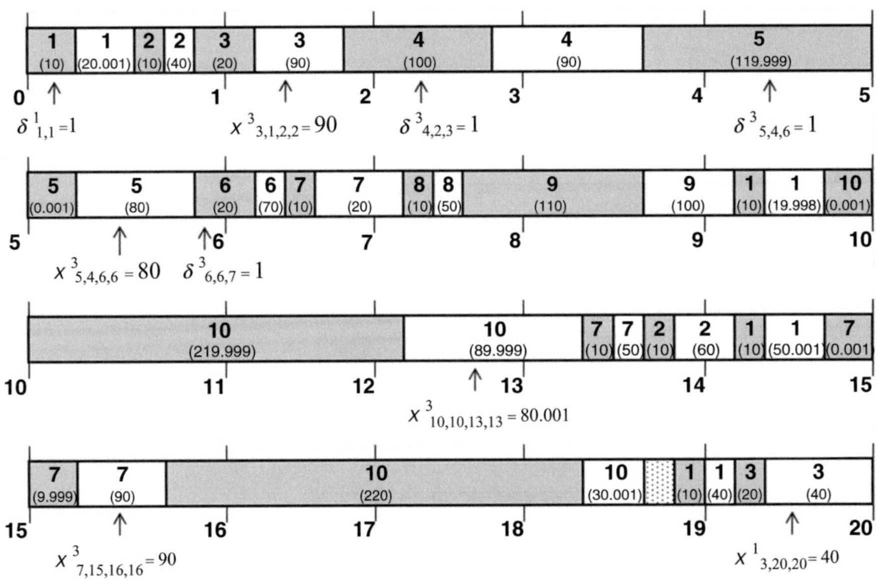

Figure 3.13: Gantt chart for the solution obtained by the proposed heuristic considering ten products and twenty time periods, when the setup costs and holding costs are product dependent and time dependent, with the setup cost for a product calculated with respect to the period of its setup initiation (for the data given in Table 3.6) considering the 2nd set of backorder costs; $Z = 4170.12$ mu

Table 3.7: Data for twelve products and twenty time periods, with two different sets of backorder costs. The Gantt charts corresponding to the exact and heuristic solutions considering the 1st set of backorder costs are given in Figs. 3.14 and 3.15; and the Gantt charts corresponding to the 2nd set of backorder costs are given in Figs. 3.16 and 3.17

Product	Setup time (ST_i): (time units)	1st set of backorder costs (b_i): (mu/period/ unit product carried over)	2nd set of backorder costs (b_i): (mu/period/ unit product carried over)	Number of time units required for producing one unit of product i (a_i): (time units/ unit of product i)
1	10	2000	20	1
2	40	3000	30	1
3	10	1000	10	1
4	20	1000	10	1
5	20	2000	20	1
6	120	4000	40	1
7	20	3000	30	1
8	100	1000	10	1
9	20	2000	20	1
10	130	4000	40	1
11	210	2000	20	1
12	40	1000	10	1

Demand ($d_{i,t}$), Setup cost ($SC_{i,t}$): (mu/setup) and Holding cost ($h_{i,t}$): (mu/period/unit product carried over)

Product \ Period	1	2	3	4	5	6	7	8	9	10	11	12	13	14	15	16	17	18	19	20
1	70	0	0	0	0	0	0	0	0	0	0	0	0	0	0	0	0	0	0	0
	20	30	10	30	50	20	40	20	40	30	20	10	5	60	30	20	30	10	25	40
	1	2	3	1	2	1	2	1	3	2	1	3	2	1	2	1	1	3	2	1
2	0	20	0	0	0	0	0	0	0	0	0	0	0	0	0	0	30	0	0	0
	40	20	40	30	20	10	5	60	30	20	20	30	10	30	50	60	60	30	20	20
	1	3	2	1	3	2	1	2	1	1	2	3	1	2	1	2	2	3	1	2
3	0	30	0	0	0	0	0	0	0	0	0	0	0	0	0	0	20	100	0	0
	30	50	60	40	20	40	30	20	20	30	15	10	5	60	30	20	20	40	30	20
	1	2	1	3	2	1	3	2	1	2	1	1	2	3	1	2	1	1	2	3

(continued)

Table 3.7: (continued)

Product \ Period	Demand ($d_{i,t}$), Setup cost ($SC_{i,t}$): (mu/setup) and Holding cost ($h_{i,t}$): (mu/period/unit product carried over)																			
	1	2	3	4	5	6	7	8	9	10	11	12	13	14	15	16	17	18	19	20
	0	0	40	0	0	0	0	0	0	0	0	0	0	0	0	0	0	0	0	0
4	20	40	30	20	10	5	60	30	20	20	30	10	30	50	60	40	30	50	60	40
	1	2	1	2	1	3	2	1	2	3	1	3	2	1	2	1	1	3	2	1
	0	0	20	0	0	0	0	0	0	0	0	0	0	0	0	40	0	0	0	0
5	60	30	20	20	30	15	30	50	60	40	20	40	30	20	10	5	40	20	40	30
	1	2	1	1	3	2	1	2	3	1	2	1	2	1	3	2	2	1	2	1
	0	0	0	0	40	0	0	0	0	0	0	0	0	0	0	0	0	0	0	0
6	30	10	25	40	50	45	60	70	15	20	35	55	10	30	50	60	20	30	10	30
	2	1	1	2	3	1	2	1	2	2	3	1	2	1	2	1	2	1	3	2
	0	0	0	0	20	0	0	0	0	0	0	0	0	0	0	0	0	0	0	0
7	20	40	30	20	20	30	15	10	5	60	30	20	20	40	20	30	10	30	50	20
	1	3	2	1	2	1	1	2	3	1	2	1	1	2	3	2	2	3	1	2
	0	0	0	0	0	0	80	0	0	0	0	0	0	0	0	0	100	0	0	0
8	20	20	30	15	30	50	60	40	20	40	30	20	10	5	40	20	40	20	30	15
	1	2	1	2	1	3	2	1	2	3	1	3	2	2	1	3	2	2	1	2

Demand ($d_{i,t}$), Setup cost ($SC_{i,t}$): (mu/setup) and Holding cost ($h_{i,t}$): (mu/period/unit product carried over)																				
Product \ Period	1	2	3	4	5	6	7	8	9	10	11	12	13	14	15	16	17	18	19	20
	0	0	0	0	0	0	0	40	0	0	0	0	0	0	0	0	0	0	0	0
9	20	30	15	30	50	60	40	20	40	30	20	10	5	40	50	20	40	20	40	30
	2	1	3	2	1	2	1	2	1	2	2	1	2	1	2	1	3	2	2	1
	0	0	0	60	0	30	0	0	0	10	0	30	0	0	60	0	0	0	0	0
10	20	40	30	20	10	5	60	30	20	20	30	10	20	20	30	15	30	50	60	40
	3	1	1	2	3	2	3	2	2	1	2	3	1	2	1	3	2	1	3	2
	0	0	0	0	0	0	0	0	0	0	20	10	0	0	0	0	0	0	0	0
11	40	30	20	20	60	15	5	20	40	20	30	20	30	20	40	20	60	40	30	20
	3	1	1	2	3	2	3	2	1	2	2	1	2	3	1	2	1	3	2	1
	0	0	0	0	0	0	0	0	0	0	0	0	0	0	0	30	0	0	0	0
12	5	10	5	30	30	10	20	10	5	20	30	60	30	60	50	20	30	40	5	40
	1	2	1	1	2	3	2	1	2	2	3	3	1	3	1	1	1	3	1	3
C_t	100	100	100	100	100	100	100	100	100	100	100	100	100	100	100	100	100	100	100	100

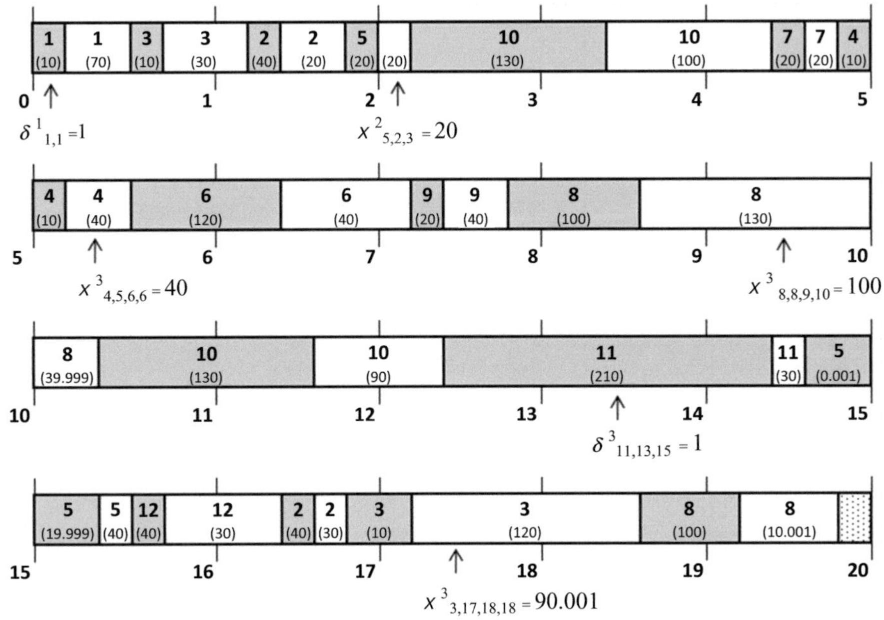

Figure 3.14: Gantt chart for the solution obtained by the proposed model MM1:CLSP-PCSC considering twelve products and twenty time periods, when the setup costs and holding costs are product dependent and time dependent, with the setup cost for a product calculated with respect to the period of its setup initiation (for the data given in Table 3.7) considering the 1st set of backorder costs; $Z = 1062095.99$ mu

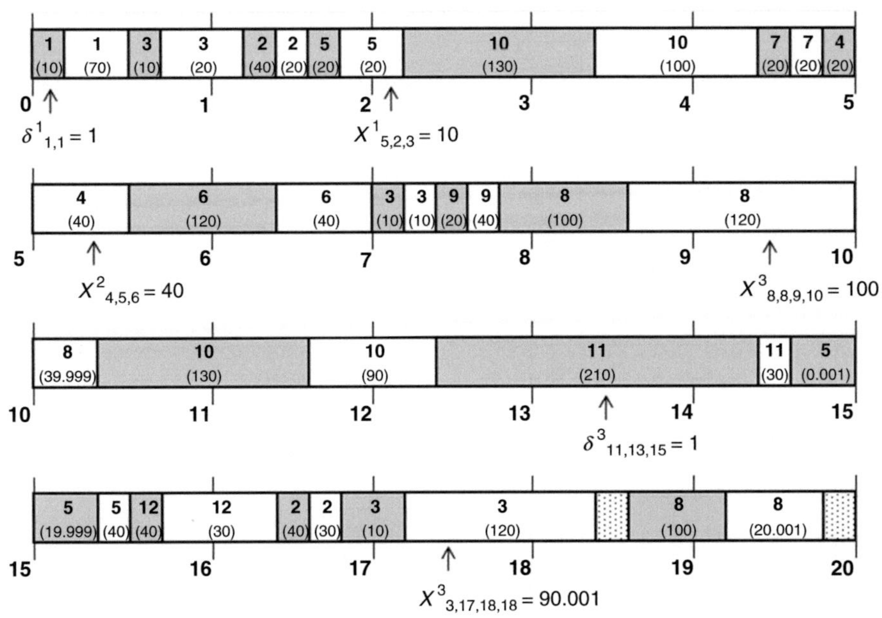

Figure 3.15: Gantt chart for the solution obtained by the proposed heuristic considering twelve products and twenty time periods, when the setup costs and holding costs are product dependent and time dependent, with the setup cost for a product calculated with respect to the period of its setup initiation (for the data given in Table 3.7) considering the 1st set of backorder costs; $Z = 1081985.99$ mu

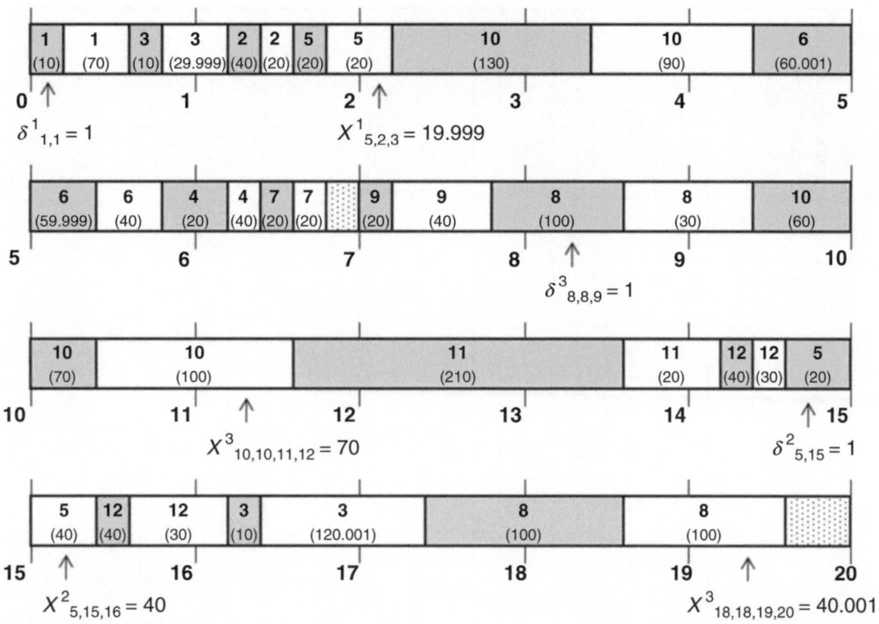

Figure 3.16: Gantt chart for the solution obtained by the proposed model MM1:CLSP-PCSC considering twelve products and twenty time periods, when the setup costs and holding costs are product dependent and time dependent, with the setup cost for a product calculated with respect to the period of its setup initiation (for the data given in Table 3.7) considering the 2nd set of backorder costs; $Z = 12490.12$ mu

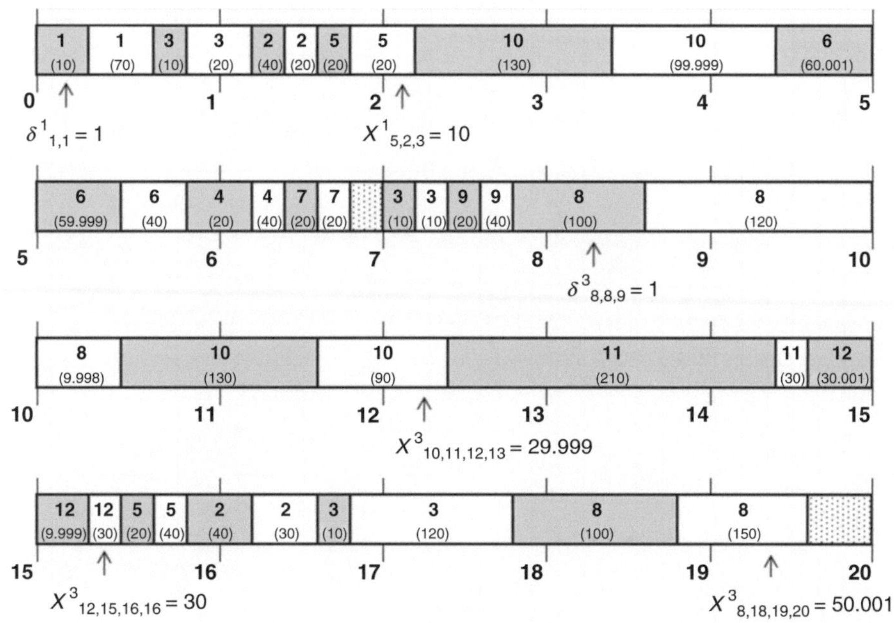

Figure 3.17: Gantt chart for the solution obtained by the proposed heuristic considering twelve products and twenty time periods, when the setup costs and holding costs are product dependent and time dependent, with the setup cost for a product calculated with respect to the period of its setup initiation (for the data given in Table 3.7) considering the 2nd set of backorder costs; $Z = 12625.10$ mu

Table 3.8: Data for fourteen products and twenty time periods, with two different sets of backorder costs. The Gantt charts corresponding to the exact and heuristic solutions considering the 1st set of backorder costs are given in Figs. 3.18 and 3.19; and the Gantt charts corresponding to the 2nd set of backorder costs are given in Figs. 3.20 and 3.21

Product	Setup time (ST_i): (time units)	1st set of backorder costs (b_i): (mu/period/ unit product carried over)	2nd set of backorder costs (b_i): (mu/period/ unit product carried over)	Number of time units required for producing one unit of product i (a_i): (time units/ unit of product i)
1	100	2000	20	1
2	10	3000	30	1
3	10	1000	10	1
4	230	1000	10	1
5	40	2000	20	1
6	50	4000	40	1
7	10	3000	30	1
8	100	1000	10	1
9	10	2000	20	1
10	20	4000	40	1
11	40	2000	20	1
12	10	1000	10	1
13	100	3000	30	1
14	110	2000	20	1

Demand ($d_{i,t}$), Setup cost ($SC_{i,t}$): (mu/setup) and Holding cost ($h_{i,t}$): (mu/period/unit product carried over)

Product \ Period	1	2	3	4	5	6	7	8	9	10	11	12	13	14	15	16	17	18	19	20
1	0	50	0	0	0	0	0	0	0	0	0	0	0	0	0	0	0	0	0	0
	20	30	10	30	50	20	40	20	40	30	20	10	5	60	30	20	30	10	25	40
	1	2	3	1	2	1	2	1	3	2	1	3	2	1	2	1	1	3	2	1
2	0	0	40	0	0	0	0	0	0	0	0	0	0	0	0	0	0	0	0	0
	40	20	40	30	20	10	5	60	30	20	20	30	10	30	50	60	60	30	20	20
	1	3	2	1	3	2	1	2	1	1	2	3	1	2	1	2	2	3	1	2
3	0	0	0	80	0	0	0	0	0	0	0	0	0	0	0	0	0	0	0	0
	30	50	60	40	20	40	30	20	20	30	15	10	5	60	30	20	20	40	30	20
	1	2	1	3	2	1	3	2	1	2	1	1	2	3	1	2	1	1	2	3

Demand ($d_{i,t}$), Setup cost ($SC_{i,t}$): (mu/setup) and Holding cost ($h_{i,t}$): (mu/period/unit product carried over)

Product \ Period	1	2	3	4	5	6	7	8	9	10	11	12	13	14	15	16	17	18	19	20
4	0	0	0	10	20	20	10	0	0	0	0	0	0	0	0	0	0	0	0	0
	20	40	30	20	10	5	60	30	20	20	30	10	30	50	60	40	30	50	60	40
	1	2	1	2	1	3	2	1	2	3	1	3	2	1	2	1	1	3	2	1
5	0	0	0	0	0	0	0	60	0	0	0	0	0	0	0	0	0	0	0	0
	60	30	20	20	30	15	30	50	60	40	20	40	30	20	10	5	40	20	40	30
	1	2	1	1	3	2	1	2	3	1	2	1	2	1	3	2	2	1	2	1
6	0	0	0	0	0	0	0	0	60	0	0	0	0	0	0	0	0	0	0	0
	30	10	25	40	50	45	60	70	15	20	35	55	10	30	50	60	20	30	10	30
	2	1	1	2	3	1	2	1	2	2	3	1	2	1	2	1	2	1	3	2
7	0	0	0	0	0	0	0	0	0	100	100	0	0	0	0	0	0	0	0	0
	20	40	30	20	20	30	15	10	5	60	30	20	20	40	20	30	10	30	50	20
	1	3	2	1	2	1	1	2	3	1	2	1	1	2	3	2	2	3	1	2
8	0	0	0	0	0	0	0	0	0	50	0	0	0	0	0	0	0	0	0	0
	20	20	30	15	30	50	60	40	20	40	30	20	10	5	40	20	40	20	30	15
	1	2	1	2	1	3	2	1	2	3	1	3	2	2	1	3	2	2	1	2
9	0	0	0	0	0	0	0	0	0	0	0	20	40	0	0	0	0	0	0	0
	20	30	15	30	50	60	40	20	40	30	20	10	5	40	50	20	40	20	40	30
	2	1	3	2	1	2	1	2	1	2	2	1	2	1	2	1	3	2	2	1

(continued)

Table 3.8: (continued)

Product \ Period	1	2	3	4	5	6	7	8	9	10	11	12	13	14	15	16	17	18	19	20
	Demand ($d_{i,t}$), Setup cost ($SC_{i,t}$): (mu/setup) and Holding cost ($h_{i,t}$): (mu/period/unit product carried over)																			
10	0	0	0	0	0	0	0	0	0	0	0	20	0	0	0	0	0	0	0	0
	20	40	30	20	10	5	60	30	20	20	30	10	20	20	30	15	30	50	60	40
	3	1	1	2	3	2	3	2	2	1	2	3	1	2	1	3	2	1	3	2
11	0	0	0	0	0	0	0	0	0	0	0	0	0	0	50	0	0	0	0	0
	40	30	20	20	60	15	5	20	40	20	30	20	30	20	40	20	60	40	30	20
	3	1	1	2	3	2	3	2	1	2	2	1	2	3	1	2	1	3	2	1
12	0	0	0	0	0	0	0	0	0	0	0	0	0	0	40	0	0	0	0	0
	5	10	5	30	30	10	20	10	5	20	30	60	30	60	50	20	30	40	5	40
	1	2	1	1	2	3	2	1	2	2	3	3	1	3	1	1	1	3	1	3
13	0	0	0	0	0	0	0	0	0	0	80	0	0	0	0	0	0	0	0	0
	10	30	40	50	20	30	30	20	20	50	40	40	20	10	40	5	15	45	30	50
	3	3	1	3	1	1	1	3	1	3	1	3	1	2	3	1	1	1	2	2
14	0	0	0	0	0	0	0	0	0	0	0	0	0	0	0	0	0	50	50	100
	30	20	50	40	30	10	40	20	30	40	30	45	15	5	40	10	20	20	5	10
	2	1	2	2	2	2	1	2	2	1	2	1	2	1	1	2	3	2	3	1
C_t	100	100	100	100	100	100	100	100	100	100	100	100	100	100	100	100	100	100	100	100

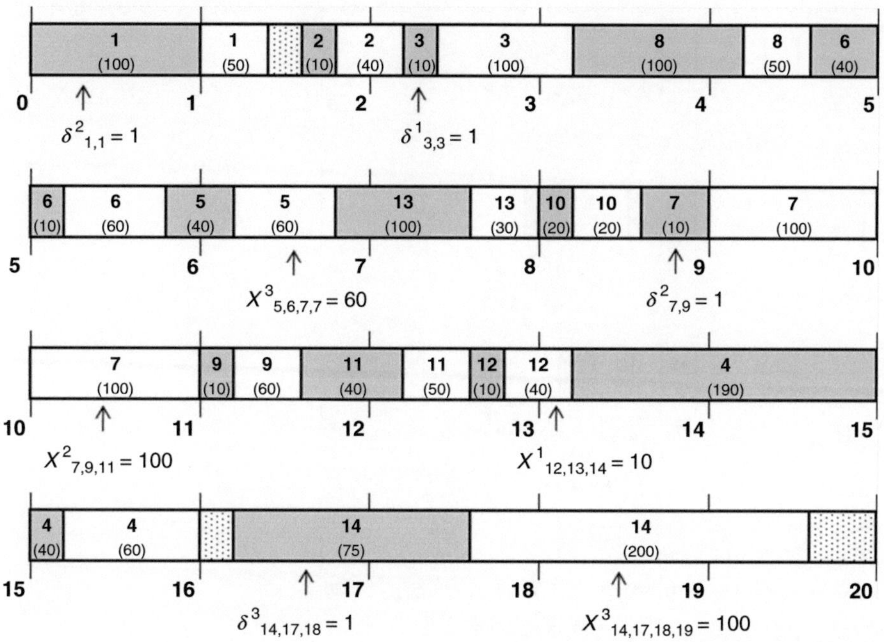

Figure 3.18: Gantt chart for the solution obtained by the proposed model MM1:CLSP-PCSC considering fourteen products and twenty time periods, when the setup costs and holding costs are product dependent and time dependent, with the setup cost for a product calculated with respect to the period of its setup initiation (for the data given in Table 3.8) considering the 1st set of backorder costs; $Z = 632345.00$ mu

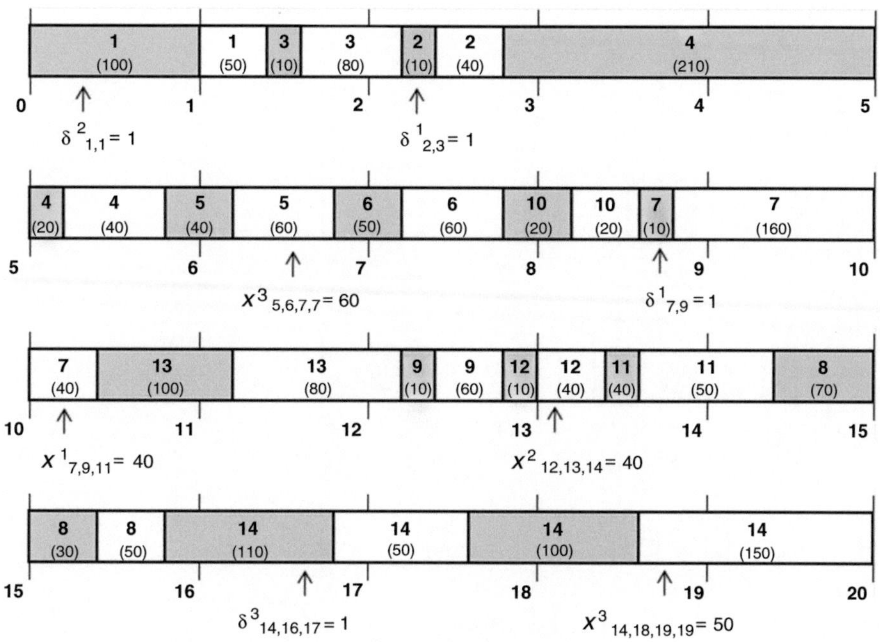

Figure 3.19: Gantt chart for the solution obtained by the proposed heuristic considering fourteen products and twenty time periods, when the setup costs and holding costs are product dependent and time dependent, with the setup cost for a product calculated with respect to the period of its setup initiation (for the data given in Table 3.8) considering the 1st set of backorder costs; $Z = 681275.00$ mu

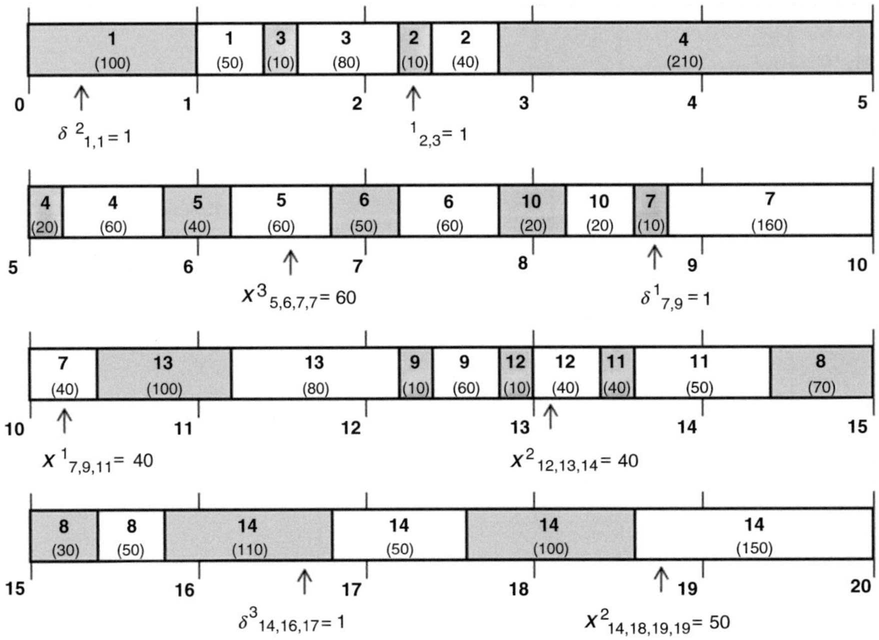

Figure 3.20: Gantt chart for the solution obtained by the proposed model MM1:CLSP-PCSC considering fourteen products and twenty time periods, when the setup costs and holding costs are product dependent and time dependent, with the setup cost for a product calculated with respect to the period of its setup initiation (for the data given in Table 3.8) considering the 2nd set of backorder costs; $Z = 8075.00$ mu

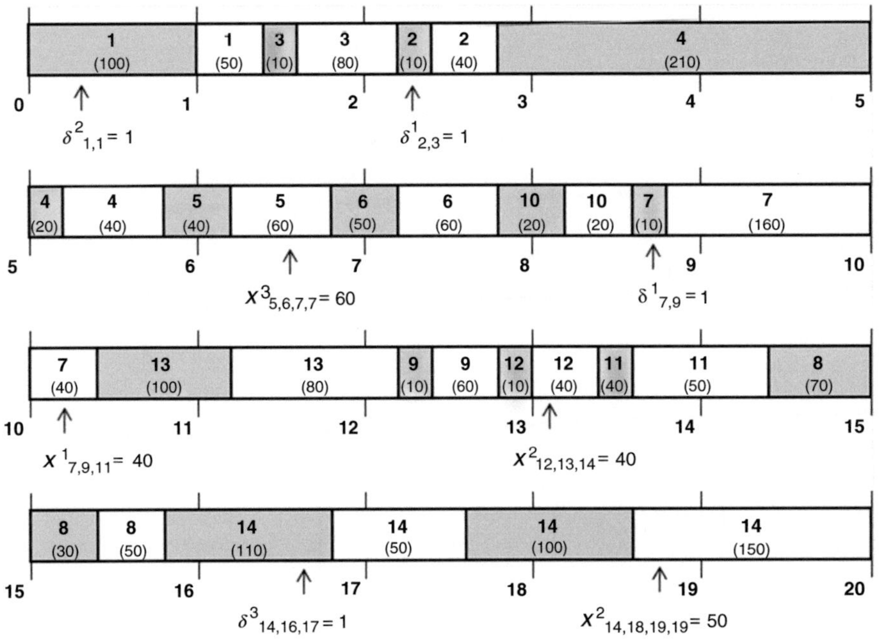

Figure 3.21: Gantt chart for the solution obtained by the proposed heuristic considering fourteen products and twenty time periods, when the setup costs and holding costs are product dependent and time dependent, with the setup cost for a product calculated with respect to the period of its setup initiation (for the data given in Table 3.8) considering the 2nd set of backorder costs; $Z = 8075.00$ mu

Table 3.9: Data for sixteen products and twenty time periods, with two different sets of backorder costs. The Gantt charts corresponding to the exact and heuristic solutions considering the 1st set of backorder costs are given in Figs. 3.22 and 3.23; and the Gantt charts corresponding to the 2nd set of backorder costs are given in Figs. 3.24 and 3.25

Product	Setup time (ST_i): (time units)	1st set of backorder costs (b_i): (mu/period/ unit product carried over)	2nd set of backorder costs (b_i): (mu/period/ unit product carried over)	Number of time units required for producing one unit of product i (a_i): (time units/ unit of product i)
1	10	2000	20	1
2	10	3000	30	1
3	100	1000	10	1
4	10	1000	10	1
5	10	2000	20	1
6	230	4000	40	1
7	40	3000	30	1
8	50	1000	10	1
9	10	2000	20	1
10	100	4000	40	1
11	10	2000	20	1
12	20	1000	10	1
13	40	3000	30	1
14	10	2000	20	1
15	100	4000	40	1
16	110	1000	10	1

Demand ($d_{i,t}$), Setup cost ($SC_{i,t}$): (mu/setup) and Holding cost ($h_{i,t}$): (mu/period/unit product carried over)

Product \ Period	1	2	3	4	5	6	7	8	9	10	11	12	13	14	15	16	17	18	19	20
	30	0	0	0	0	0	0	0	0	0	0	0	0	0	0	0	0	0	0	0
1	20	30	10	30	50	20	40	20	40	30	20	10	5	60	30	20	30	10	25	40
	1	2	3	1	2	1	2	1	3	2	1	3	2	1	2	1	1	3	2	1
	50	0	0	0	0	0	0	0	0	0	0	0	0	0	0	0	0	0	0	0
2	40	20	40	30	20	10	5	60	30	20	20	30	10	30	50	60	60	30	20	20
	1	3	2	1	3	2	1	2	1	1	2	3	1	2	1	2	2	3	1	2
	0	50	0	0	0	0	0	0	0	0	0	0	0	0	0	0	0	0	0	0
3	30	50	60	40	20	40	30	20	20	30	15	10	5	60	30	20	20	40	30	20
	1	2	1	3	2	1	3	2	1	2	1	1	2	3	1	2	1	1	2	3

(continued)

Table 3.9: (continued)

Product \ Period	Demand ($d_{i,t}$), Setup cost ($SC_{i,t}$): (mu/setup) and Holding cost ($h_{i,t}$): (mu/period/unit product carried over)																			
	1	2	3	4	5	6	7	8	9	10	11	12	13	14	15	16	17	18	19	20
	0	0	40	0	0	0	0	0	0	0	0	0	0	0	0	0	0	0	0	0
4	20	40	30	20	10	5	60	30	20	20	30	10	30	50	60	40	30	50	60	40
	1	2	1	2	1	3	2	1	2	3	1	3	2	1	2	1	1	3	2	1
	0	0	0	80	0	0	0	0	0	0	0	0	0	0	0	0	0	0	0	0
5	60	30	20	20	30	15	30	50	60	40	20	40	30	20	10	5	40	20	40	30
	1	2	1	1	3	2	1	2	3	1	2	1	2	1	3	2	2	1	2	1
	0	0	0	10	20	20	10	0	0	0	0	0	0	0	0	0	0	0	0	0
6	30	10	25	40	50	45	60	70	15	20	35	55	10	30	50	60	20	30	10	30
	2	1	1	2	3	1	2	1	2	2	3	1	2	1	2	1	2	1	3	2
	0	0	0	0	0	0	0	60	0	0	0	0	0	0	0	0	0	0	0	0
7	20	40	30	20	20	30	15	10	5	60	30	20	20	40	20	30	10	30	50	20
	1	3	2	1	2	1	1	2	3	1	2	1	1	2	3	2	2	3	1	2
	0	0	0	0	0	0	0	0	60	0	0	0	0	0	0	0	0	0	0	0
8	20	20	30	15	30	50	60	40	20	40	30	20	10	5	40	20	40	20	30	15
	1	2	1	2	1	3	2	1	2	3	1	3	2	2	1	3	2	2	1	2
	0	0	0	0	0	0	0	0	0	100	100	0	0	0	0	0	0	0	0	0
9	20	30	15	30	50	60	40	20	40	30	20	10	5	40	50	20	40	20	40	30
	2	1	3	2	1	2	1	2	1	2	2	1	2	1	2	1	3	2	2	1
	0	0	0	0	0	0	0	0	0	50	0	0	0	0	0	0	0	0	0	0
10	20	40	30	20	10	5	60	30	20	20	30	10	20	20	30	15	30	50	60	40
	3	1	1	2	3	2	3	2	2	1	2	3	1	2	1	3	2	1	3	2

Demand ($d_{i,t}$), Setup cost ($SC_{i,t}$): (mu/setup) and Holding cost ($h_{i,t}$): (mu/period/unit product carried over)																				
Product \ Period	1	2	3	4	5	6	7	8	9	10	11	12	13	14	15	16	17	18	19	20
	0	0	0	0	0	0	0	0	0	0	0	20	40	0	0	0	0	0	0	0
11	40	30	20	20	60	15	5	20	40	20	30	20	30	20	40	20	60	40	30	20
	3	1	1	2	3	2	3	2	1	2	2	1	2	3	1	2	1	3	2	1
	0	0	0	0	0	0	0	0	0	0	0	20	0	0	0	0	0	0	0	0
12	5	10	5	30	30	10	20	10	5	20	30	60	30	60	50	20	30	40	5	40
	1	2	1	1	2	3	2	1	2	2	3	3	1	3	1	1	1	3	1	3
	0	0	0	0	0	0	0	0	0	0	0	0	0	0	50	0	0	0	0	0
13	10	30	40	50	20	30	30	20	20	50	40	40	20	10	40	5	15	45	30	50
	3	3	1	3	1	1	1	3	1	3	1	3	1	2	3	1	1	1	2	2
	0	0	0	0	0	0	0	0	0	0	0	0	0	0	40	0	0	0	0	0
14	30	20	50	40	30	10	40	20	30	40	30	45	15	5	40	10	20	20	5	10
	2	1	2	2	2	2	1	2	2	1	2	1	2	1	1	2	3	2	3	1
	0	0	0	0	0	0	0	0	0	0	80	0	0	0	0	0	0	0	0	0
15	50	30	10	20	20	50	40	10	20	20	5	20	60	60	20	40	30	20	60	30
	1	2	3	1	1	2	2	2	1	2	1	3	2	1	1	2	2	1	2	3
	0	0	0	0	0	0	0	0	0	0	0	0	0	0	0	0	0	50	50	100
16	45	15	5	40	10	20	20	5	10	60	10	15	5	10	30	20	50	40	30	10
	2	1	1	2	3	1	1	1	2	1	2	4	1	2	1	1	1	1	1	2
C_t	100	100	100	100	100	100	100	100	100	100	100	100	100	100	100	100	100	100	100	100

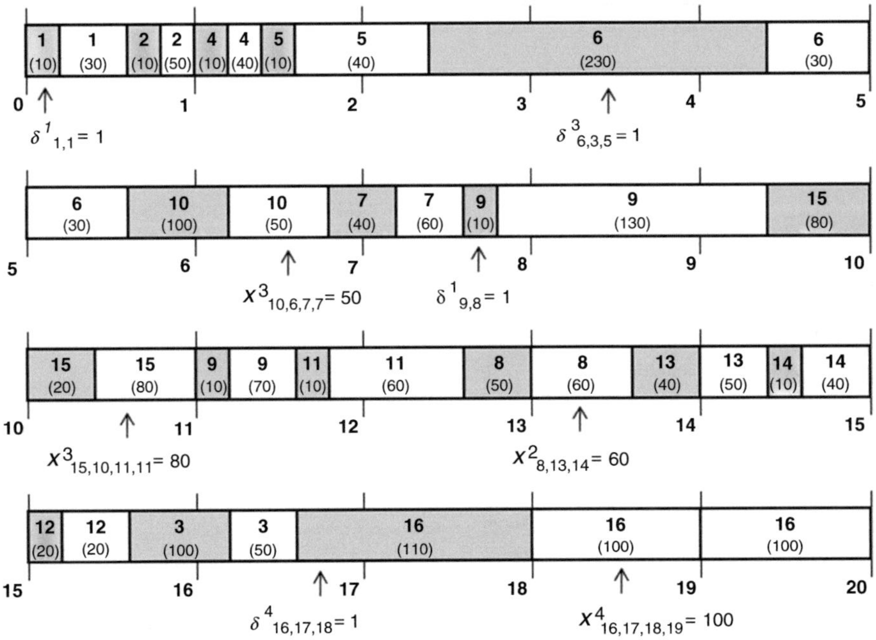

Figure 3.22: Gantt chart for the solution obtained by the proposed model MM1:CLSP-PCSC considering sixteen products and twenty time periods, when the setup costs and holding costs are product dependent and time dependent, with the setup cost for a product calculated with respect to the period of its setup initiation (for the data given in Table 3.9) considering the 1st set of backorder costs; $Z = 1381185.00$ mu

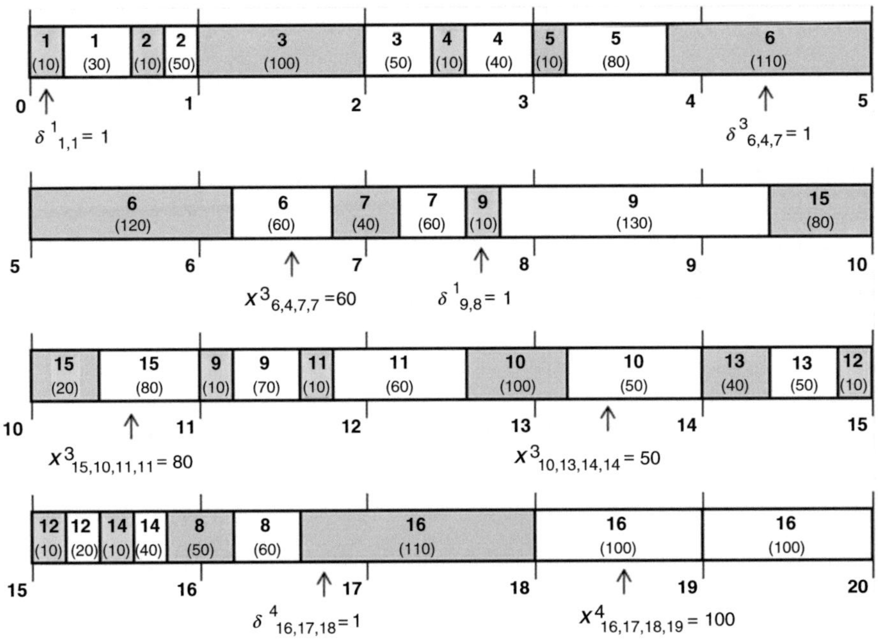

Figure 3.23: Gantt chart for the solution obtained by the proposed heuristic considering sixteen products and twenty time periods, when the setup costs and holding costs are product dependent and time dependent, with the setup cost for a product calculated with respect to the period of its setup initiation (for the data given in Table 3.9) considering the 1st set of backorder costs; $Z = 2060665.00$ mu

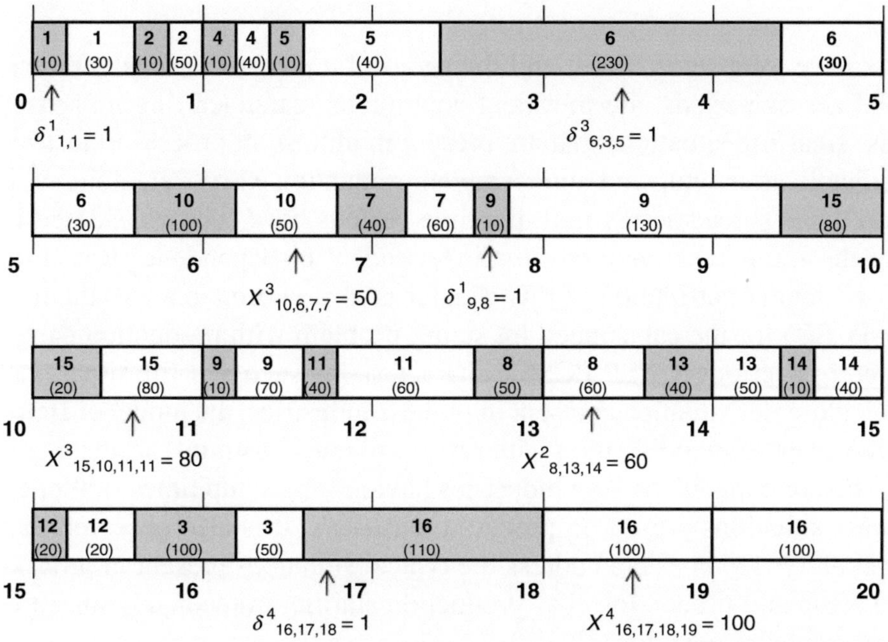

Figure 3.24: Gantt chart for the solution obtained by the proposed model MM1:CLSP-PCSC considering sixteen products and twenty time periods, when the setup costs and holding costs are product dependent and time dependent, with the setup cost for a product calculated with respect to the period of its setup initiation (for the data given in Table 3.9) considering the 2nd set of backorder costs; $Z = 14985.00$ mu

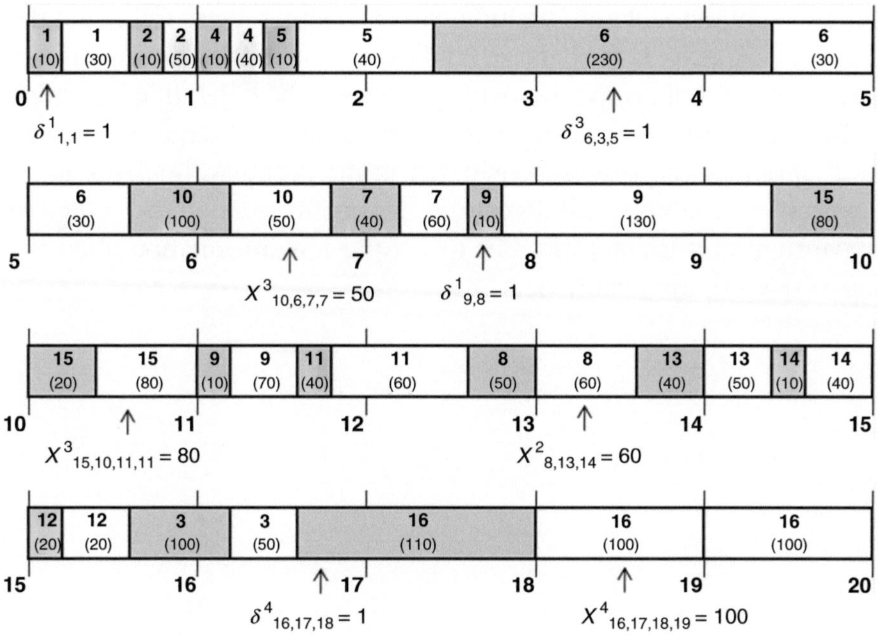

Figure 3.25: Gantt chart for the solution obtained by the proposed heuristic considering sixteen products and twenty time periods, when the setup costs and holding costs are product dependent and time dependent, with the setup cost for a product calculated with respect to the period of its setup initiation (for the data given in Table 3.9) considering the 2nd set of backorder costs; $Z = 21265.00$ mu

3.8 Summary

Situations where products have long setup times, and the setup of a product and its consecutive production are carried across periods, are prevalent continuous manufacturing industries such as process industries. Real-life situations that are present in almost all process industries such as production immediately after setup and uninterrupted production carryover, along with the presence of long setup times spread across multiple time periods have been considered in this chapter. In addition, the setup costs may be time independent or dependent. Based on these real-life situations, the main contribution of this chapter is to propose a new Mathematical Model (MM1:CLSP-PCSC) for the capacitated lot sizing problem with production carryover and setup crossover across periods (CLSP-PCSC), characterizing such real-life situations. While the proposed model addresses situations present in process industries, the model of Belo-Filho et al. (2013) (with the proposed modification with respect to one constraint) seems to address situations present in discrete manufacturing industries having long setup times (although the authors claim to address situations present in process industries). It is also observed that the model by Belo-Filho et al. (2013) does not address the typical situations present in process industries with respect to setup and production (i.e. production starting immediately after the product is set up and uninterrupted production). It is also evident that the approach used in proposed mathematical model is completely different from that of the existing models. Along with the mathematical models, a heuristic has also been proposed (based on MM1:CLSP-PCSC) for solving CLSP-PCSC, CLSP being an NP-hard problem (see Bitran and Yanasse (1982)). From the computational experience, the mathematical model seems effective from the point of view of both the solution quality and the computational time. Problems of various sizes which were executed to show the computational experience seem computationally fast. The computational effort does depend upon the capacity available versus demand of the products across the planning horizon. When the initial computational investigation was carried out it was found out that when there is less demand and the capacity is fairly large (and hence resulting in the idle time of the machine resource), there is no need of setup crossover. Therefore, the CPU time was much less in comparison to the situation where the capacity availability and demand requirement were fairly tight. However, "hard" problems have been considered in the proposed heuristic and it is hence ensured that the capacity availability and demand requirement were almost matching in every period (and we have tight constraints) in order to get the fair information about the computational performance of the proposed heuristic.

CHAPTER 4

Further Development: Mathematical Model 2 (MM2) and a Comprehensive Heuristic for Capacitated Lot Sizing Problem with Production Carryover and Setup Crossover Across Periods for Process Industries

4.1 Introduction and Problem Definition

In the previous chapter, a mathematical model and a heuristic are applied to the CLSP in process industries which can be applied to real-life situations in process industries such as production carryover across periods and setup crossover across periods. The heuristic proposed in Chap. 3 with respect to MM1:CLSP-PCSC can be easily applied when identical capacity is present across periods. However, in reality the capacity across periods may be varying. When non-identical capacity is present across periods, for allowing shift of setup/production for more periods ahead of or after the current time period, the extension of the heuristic based on MM1:CLSP-PCSC becomes tedious. In such cases the heuristic proposed in this chapter is easier to apply. Hence, in this chapter we propose a second mathematical model (MM2:CLSP-PCSC) for the CLSP-PCSC followed by a heuristic using the second mathematical model. The proposed model in this chapter is not constrained by the consideration of long setup products. The model is flexible enough to handle the process industries with small bucket setups and long bucket production runs or the scenario with large bucket setups and small production runs or a mixture of both. In other words, the proposed mathematical model and heuristic approach are flexible enough to handle or address situations in the conventional process industries such as cement and sugar industries (associated with small bucket setups and long bucket production runs), large bucket setups and small bucket production runs (associated with highly technological intensive big bucket setups and small bucket production runs such as those in highly specialized pharmaceutical processes), or a mixture of scenarios in a single process industry. Also, depending upon the industry the definition of a period may vary. It is to be noted that in all these scenarios we have real-life restrictions that once a process starts there is no interruption with the production run length, and the production has to start immediately after the completion of setup. In this book we address such a variety or mix of process-industry scenarios and the restriction in terms of continuous production and production commencement immediately after setup. This book is primarily motivated by the literature on CLSP based on the nature of continuous manufacturing industries such as chemical manufacturing, cement manufacturing, sugar industries, pharmaceuticals, hot rolling process, heat treatment, casting and injection moulding, and a real-life case study in a batch processing industry. Referring to the benchmark literature (e.g. Sung and Maravelias (2008) and Belo-Filho et al. (2013)),

R. Ramya et al., *Capacitated Lot Sizing Problems in Process Industries*, https://doi.org/10.1007/978-3-030-01222-9_4

we find that no existing work has attempted such a mix of industrial scenarios and associated real-life constraints such as continuous production with no interruption and production commencement immediately after setup completion. Therefore, the proposed mathematical model in this chapter is also generalized in nature.

Similar to Chap. 3, there can be three ways by which a machine is set up for producing a product. They are: (i) a machine is completely setup for product i anywhere in period t, and the production starts in period t itself and the production may be continued to the period(s) thereafter; (ii) when there is enough capacity left at the end of a period, it can be utilized in making a complete setup for a product, followed by its production in time period $t + 1$. An end-of-period setup of product i can occur when an amount of capacity is left at the end of period t to setup the product, or it can also refer to the setup which takes a value equal to the capacity of the period in which its setup is initiated; and (iii) the setup started in period t can be carried over to period t', followed by its production in time period t' or $t'' > t'$ depending upon where the setup of product i ends (this aspect is referred to as setup crossover in this book). The phenomenon setup crossover also includes the concept of setup splitting between periods t and $t + 1$, when the setup time of product i is less that the capacity of period t in which its setup is initiated, and less than the capacity of period $t+1$ in which its setup is completed. The figures corresponding to the three ways of setup are shown in Fig. 4.1.

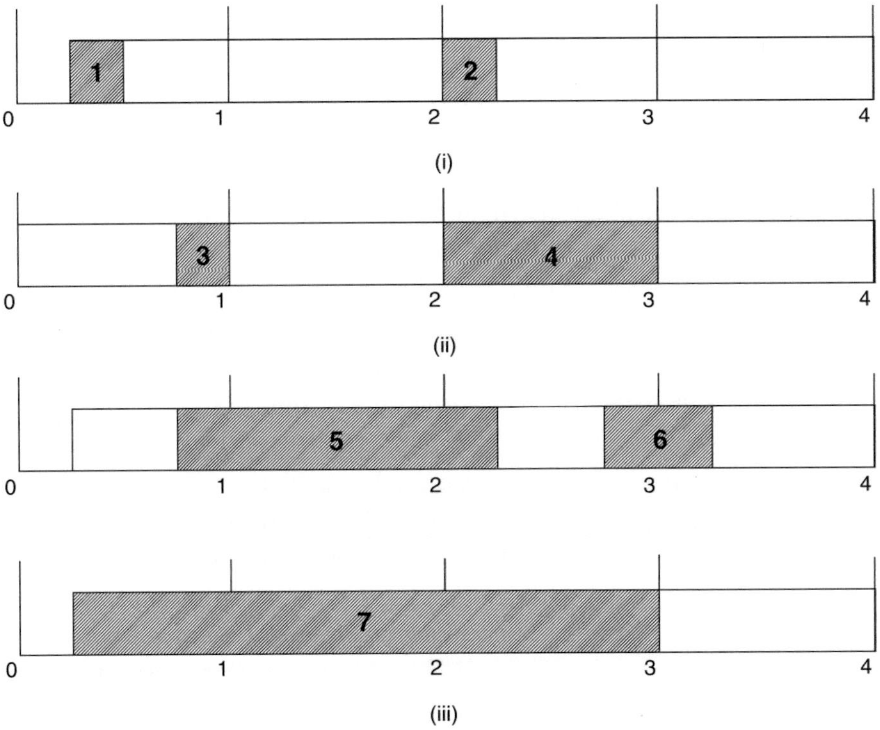

Figure 4.1: Three ways of a machine being set up in a period for production in process industries

Remark: In Fig. 4.1, the X-axis in the Gantt charts denote the time periods represented as 1, 2, 3 and so on. The shaded region indicates the setup and the blank region indicates the production. Figure (i) indicates a complete setup for product 1 in period 1, with its production starting in period 1 and continued to period 2; another complete setup for product 2 is initiated in period 3, with its production starting in period 3 and continued to period 4 (this setup situation is indicated by $\delta^1_{i,t}$ in MM2:CLSP-PCSC). Figure (ii) indicates an end-of-period setup for product

3 in period 1, with its production starting in period 2; and another end-of-period setup for product 4 in period 3, with its production starting in period 4 (this setup situation is indicated by $\delta^2_{i,t}$ in MM2:CLSP-PCSC). The first Gantt chart in (iii) indicates a setup crossover of product 5, starting in period 1 and carried over to period 3, followed by its production in time period 3; and another setup crossover of product 6, starting in period 3 and ending in period 4, followed by its production in period 4 (this setup situation is indicated by $\delta^3_{i,t}$ linked with $\Omega^3_{i,t,t}$ in MM2:CLSP-PCSC). The second Gantt chart in (iii) indicates a setup crossover of product 7, starting in period 1 and carried over to period 3, ending exactly at the end of period 3, followed by its production in time period 4 (this setup situation is indicated by $\delta^3_{i,t}$ linked with $\Omega^4_{i,t,t}$ in MM2:CLSP-PCSC).

Note: Description of all the decision variables is provided in Sect. 4.3.2.

The chapter is organized as follows. Section 4.2 presents the basic assumptions of the proposed mathematical model (MM2:CLSP-PCSC). Section 4.3 presents the parameters and decision variables for MM2:CLSP-PCSC (see the corresponding Sects. 4.3.1 and 4.3.2, respectively). The generalized version of mathematical model MM2:CLSP-PCSC considering the scenario where the setup cost of a product is calculated with respect to the period of initiation of setup of the product is presented in Sect. 4.3.3. Method of tracking setups in MM2:CLSP-PCSC is presented in Sect. 4.3.4. In Sect. 4.4, special cases of MM2:CLSP-PCSC are presented with appropriate changes made to the objective function and add-on constraints. They are: MM2:CLSP-PCSC when the setup cost of a product is calculated with respect to the period of its setup completion (Sect. 4.4.1) and MM2:CLSP-PCSC when the setup cost and holding cost of a product are time independent (Sect. 4.4.2). Following this, Sect. 4.5 presents the numerical illustration and discussion for the generalized MM2:CLSP-PCSC and its special cases. A comprehensive heuristic is proposed with respect to MM2:CLSP-PCSC, and implemented for solving the CLSP-PCSC (see Sect. 4.6). The computational time for MM2:CLSP-PCSC is presented in Sect. 4.7. The chapter is concluded with a summary of the contributions and is presented in Sect. 4.8.

4.2 Basic Assumptions of the Proposed Mathematical Model (MM2: CLSP-PCSC)

- A single machine is considered in the problem.
- Multiple products can be produced on the single machine and each product is made up of a single level.
- Time unit is discrete and the time horizon considered is finite.
- Each product is associated with a setup cost when set up on the machine, and it consumes time for setup.
- Backorders are allowed but lost sales are not permitted.
- The capacity of the machine during a given period is assumed in time units and it may vary from period to period.
- The capacity of the machine per period is consumed by the setup time and the production time of the products. Idle time on the machine can also be present.

- If excess capacity is left over on the machine in a period after production in period t, it may be used to setup the product to be produced in the next period. If the setup is not over, this setup may be continued to some future period t', where $t' > t$. In this work this aspect is called setup crossover.
- The excess quantity produced of a product can be stored and this incurs a holding cost, except in the last period where all the units in the inventory have to be consumed.
- Production of a product may extend over any number of periods subject to demand and capacity constraints. In this work, this aspect is called production carryover.
- At most one setup of a given product i can be initiated in the given time period t. It means that the carryover of a setup is permitted from one of the previous time periods to the present period, and the initiation of setup of the same product in that given time period is also permitted.

4.3 Mathematical Model 2 (MM2:CLSP-PCSC) for the Capacitated Lot Sizing Problem with Production Carryover and Setup Crossover Across Periods

The parameters, decision variables and the generalized mathematical model MM2:CLSP-PCSC are presented in this section.

4.3.1 Parameters/Indices

N	number of products
T	number of time periods
t	a given time period
i	product
$SC_{i,t}$	setup cost for product i when its setup is initiated in period t; this cost is incurred only once as a fixed cost computed with respect to the period of its setup initiation
b_i	backorder cost per period per unit of product i
$h_{i,t}$	holding cost per period per unit of product i in period t
ST_i	setup time for product i
a_i	number of time units required for producing one unit of product i
C_t	capacity of the machine in period t (in time units)
$d_{i,t}$	demand for product i in period t
M	a large value
$\mathcal{E}$	smallest unit of time
$\mathcal{E}_d$	unit of smallest quantity of production

4.3.2 Decision Variables

Variable	Description
$\delta^1_{i,t}$	An indicator (binary) variable that takes value 1 if a complete setup is done for product i in period t with the production starting in period t; 0 otherwise.
$\Delta^1_{i,t,t'}$	An indicator (binary) variable that takes value 1: it corresponds to a possible production carryover from period t to period t' ($t \leq t' \leq T$), due to the setup of product i started and finished in period t, with no intermittent setup of any other product; 0 otherwise.
$\delta^2_{i,t}$	An indicator (binary) variable that takes value 1 if the setup of product i is started and completed exactly at the end of period t followed by its production starting in period $t+1$; 0 otherwise.
$\Delta^2_{i,t,t'}$	An indicator (binary) variable that takes value 1: it corresponds to a possible production carryover from period t' to period $t'+1$ ($t+1 \leq t' \leq T$), due to the end-of-period setup of product i in period t, with no intermittent setup of any other product; 0 otherwise.
$\delta^3_{i,t}$	An indicator (binary) variable that takes value 1 if the setup of product i is commenced in period t and is carried over across periods, and is completed in some period t' ($t' = t+1, t+2, .., T$); 0 otherwise.
$\Omega^3_{i,t,t'}$	An indicator (binary) variable that takes value 1 if the setup of product i is initiated in period t and is present in period t' ($t \leq t' \leq T$), with the setup of product i ending in a period later than period t, but not exactly at the end of that period; 0 otherwise.
$\Omega^4_{i,t,t'}$	An indicator (binary) variable that takes value 1 if the setup of product i is initiated in period t and is present in period t' ($t \leq t' \leq T-1$), with the setup of product i ending in a period later than period t and setup getting completed exactly at the end of that period; 0 otherwise.
$\Delta^3_{i,t,t'}$	An indicator (binary) variable that takes value 1: it corresponds to a possible production carryover from period t' to period $t'+1$ ($t+1 \leq t' \leq T$), with the setup of product i (having started in period t) ending in a period later than period t, but not exactly at the end of that period, and with no intermittent setup of any product during the production of product i; 0 otherwise.

Variable	Description
$\Delta^4_{i,t,t'}$	An indicator (binary) variable that takes value 1: it corresponds to a possible production carryover from period t' to period $t'+1$ ($t+2 \le t' \le T$), with the setup of product i (having started in period t) ending in a period later than period t and setup getting completed exactly at the end of that period, and with no intermittent setup of any product during the production of product i; 0 otherwise.
$I_{i,t}$	Inventory of product i at the end of period t.
$B_{i,t}$	Backorder quantity of product i at the end of period t.
$s^1_{i,t}$	Setup time of product i in period t that takes the value of ST_i, and associated with $\delta^1_{i,t}$.
$s^2_{i,t}$	Setup time of product i in period t that takes the value of ST_i, and associated with $\delta^2_{i,t}$.
$s^3_{i,t,t'}$	Setup time of product i in period t' due to its setup started in period t, and associated with $\Omega^3_{i,t,t'}$.
$s^4_{i,t,t'}$	Setup time of product i in period t' due to its setup started in period t, and associated with $\Omega^4_{i,t,t'}$.
$X^1_{i,t,t'}$	Production quantity of product i in period t' (due to its setup starting and ending within period t), with $1 \le t \le T$ and $t \le t' \le T$, and associated with $\Delta^1_{i,t,t'}$.
$X^2_{i,t,t'}$	Production quantity of product i in period t' (due to its setup started in period t and completed exactly at the end of that period), with $1 \le t \le T-1$ and $t+1 \le t' \le T$, and associated with $\Delta^2_{i,t,t'}$.
$X^3_{i,t,t'}$	Production quantity of product i in period t' (due to its setup starting in period t and ending in a later period, but not at the end of that period), with $1 \le t \le T-1$ and $t+1 \le t' \le T$, and associated with $\Delta^3_{i,t,t'}$.
$X^4_{i,t,t'}$	Production quantity of product i in period t' (due to its setup starting in period t and ending at the end of a later period), with $1 \le t \le T-2$ and $t+2 \le t' \le T$, and associated with $\Delta^4_{i,t,t'}$.

4.3.3 Mathematical Model 2 (CLSP-PCSC)

In this section, the mathematical model (MM2:CLSP-PCSC) is presented, with an objective of minimizing the time-dependent setup costs, holding costs and backorder costs of all products across all time periods. The proposed mathematical model assumes that the setup cost of a product is calculated with respect to the period of its setup initiation. This mathematical model helps to address production situations in process industries. They address situations such as the presence of long setup times, production starting immediately after the product's setup and uninterrupted production carryover across periods. In this mathematical model, three binary variables are used to track three types of setup, i.e., a complete setup done (started and finished) anywhere in period t (but not exactly at the end of period t), an end-of-period setup and a setup crossover which is initiated in period t due to the presence of long setup times. This setup crossover indicator is in turn linked with two other binary variables. One which indicates a setup crossing over across a number of periods with the setup initiated anywhere in period t and ending in the middle of some future period t', and another which indicates a setup crossing over across a number of periods with the setup initiated anywhere in period t and completing exactly at the end of period t'. These binary variables which indicate a product's setup are in turn linked with production carryover indicator variables that help to track the time period in which the product's setup is initiated and completed, and the time period in which the corresponding production is carried out. Production variables corresponding to these production carryover indicator variables are also present which help to determine the production time (here, the production quantity is measured in time units). There are variables which determine the setup time of a product in a period. When the setup of product i crosses over a number of periods, the time taken to setup product i is also split across these periods. There are also constraints to ensure that production starts immediately after setup and uninterrupted production takes place across periods. Through this mathematical model, it is ensured that the demand for all products is satisfied across the entire time horizon with the condition that the production time and the setup time of the products setup in a period do not exceed the capacity limitations (measured in time units) of that period.

Objective Function:

$$\text{Min } Z = \sum_{i=1}^{N}\sum_{t=1}^{T} SC_{i,t}\,\delta^1_{i,t} + \sum_{i=1}^{N}\sum_{t=1}^{T-1} SC_{i,t}\,\delta^2_{i,t} + \sum_{i=1}^{N}\sum_{t=1}^{T-1} SC_{i,t}\,\delta^3_{i,t} + \sum_{i=1}^{N}\sum_{t=1}^{T} h_{i,t}\,I_{i,t} + \tag{4.1}$$

$$\sum_{i=1}^{N}\sum_{t=1}^{T} b_i\,B_{i,t}$$

Subject to the Following:

/* Constraints (4.2)–(4.3) represent the conditions for setting up a product only once in period t */

$$\textstyle\sum_{i=1}^{N}(\delta^2_{i,t} + \delta^3_{i,t}) \le 1, \qquad t=1,2,..,T\text{-}1. \tag{4.2}$$

$$(\delta^1_{i,t} + \delta^2_{i,t} + \delta^3_{i,t}) \le 1 \qquad \forall i \text{ and } t=1,2,..,T\text{-}1. \tag{4.3}$$

/* Constraints (4.4)–(4.10) capture a possible complete setup within period t, with the production starting in the same period t */

$$\Delta^1_{i,t,t} = \delta^1_{i,t} \qquad \forall i \text{ and } \forall t. \tag{4.4}$$

$$\Delta^1_{i,t,t'} \ge \Delta^1_{i,t,t'+1} \qquad \forall i,\ t=1,2,..,T\text{-}1 \text{ and } t'=t,t+1,..,T\text{-}1. \tag{4.5}$$

$$\textstyle\sum_{i'=1}^{N}(\delta^1_{i',t'-1}+\delta^2_{i',t'-1}+\delta^3_{i',t'-1}) \le N(1-\Delta^1_{i,t,t'}) \quad \forall i,\ t=1,2,..,T\text{-}2 \text{ and } t'=t+2,t+3,..,T. \tag{4.6}$$

$$X^1_{i,t,t'} \geq \mathcal{E}_d - \frac{C_{t'}}{a_i}(1 - \Delta^1_{i,t,t'}) \qquad \forall i,\ \forall t \text{ and } t'=t,t+1,..,T. \qquad (4.7)$$

$$a_i\, X^1_{i,t,t'} \leq C_{t'}\, \Delta^1_{i,t,t'} \qquad \forall i,\ \forall t \text{ and } t'=t,t+1,..,T. \qquad (4.8)$$

$$s^1_{i,t} = ST_i\, \delta^1_{i,t} \qquad \forall i \text{ and } \forall t. \qquad (4.9)$$

$$s^1_{i,t} \leq (C_t - \mathcal{E}) + C_t(1 - \delta^1_{i,t}) \qquad \forall i \text{ and } \forall t. \qquad (4.10)$$

/* Constraints (4.11)–(4.19) correspond to a possible end-of-period setup in period t, with the production starting in period $t+1$ */

$$\Delta^2_{i,t,t'} = \delta^2_{i,t} \qquad \forall i,\ t=1,2,..,T\text{-}1 \text{ and } t'=t+1. \qquad (4.11)$$

$$\Delta^2_{i,t,t'} \geq \Delta^2_{i,t,t'+1} \qquad \forall i,\ t=1,2,..,T\text{-}2 \text{ and } t'=t+1,t+2,..,T\text{-}1. \qquad (4.12)$$

$$\sum_{i'=1}^{N}(\delta^1_{i',t'-1} + \delta^2_{i',t'-1} + \delta^3_{i',t'-1}) \leq N(1 - \Delta^2_{i,t,t'}) \qquad \forall i,\ t=1,2,..,T\text{-}2 \text{ and } t'=t+2,t+3,..,T. \qquad (4.13)$$

$$X^2_{i,t,t'} \geq \mathcal{E}_d - \frac{C_{t'}}{a_i}(1 - \Delta^2_{i,t,t'}) \qquad \forall i,\ t=1,2,..,T\text{-}1 \text{ and } t'=t+1,t+2,..,T. \qquad (4.14)$$

$$a_i\, X^2_{i,t,t'} \leq C_{t'}\, \Delta^2_{i,t,t'} \qquad \forall i,\ t=1,2,..,T\text{-}1 \text{ and } t'=t+1,t+2,..,T. \qquad (4.15)$$

$$s^2_{i,t} = ST_i\, \delta^2_{i,t} \qquad \forall i \text{ and } t=1,2,..,T\text{-}1. \qquad (4.16)$$

$$s^2_{i,t} \leq C_t + C_t(1 - \delta^2_{i,t}) \qquad \forall i \text{ and } t=1,2,..,T\text{-}1. \qquad (4.17)$$

$$\Delta^2_{i,t,t} = X^2_{i,t,t} = 0 \qquad \forall i \text{ and } \forall t. \qquad (4.18)$$

$$\delta^2_{i,T} = 0 \qquad \forall i. \qquad (4.19)$$

/* Constraints (4.20)–(4.39) represent a possible setup crossover across a number of periods with the production starting in the same period where the setup ends */

$$\Omega^3_{i,t,t} \leq \delta^3_{i,t} \quad \forall i \text{ and } t\text{=1,2,..,}T\text{-1.} \tag{4.20}$$

$$\Omega^3_{i,t,t+1} = \Omega^3_{i,t,t} \quad \forall i \text{ and } t\text{=1,2,..,}T\text{-1.} \tag{4.21}$$

$$\Omega^3_{i,t,t'} \geq \Omega^3_{i,t,t'+1} \quad \forall i, t\text{=1,2,..,}T\text{-1 and } t'\text{=}t,t\text{+1,..,}T\text{-1.} \tag{4.22}$$

$$\Omega^3_{i,t,t'} + \Delta^3_{i,t,t'} \geq \Delta^3_{i,t,t'+1} \quad \forall i, t\text{=1,2,..,}T\text{-1 and } t'\text{=}t,t\text{+1,..,}T. \tag{4.23}$$

$$\Delta^3_{i,t,t'} \leq \Omega^3_{i,t,t} \quad \forall i, t\text{=1,2,..,}T\text{-1 and } t'\text{=}t,t\text{+1,..,}T. \tag{4.24}$$

$$\Omega^3_{i,t,t'+1} + \Delta^3_{i,t,t'} \leq 1 \quad \forall i, t\text{=1,2,..,}T\text{-1 and } t'\text{=}t\text{+1,}t\text{+2,..,}T. \tag{4.25}$$

$$\Omega^3_{i,t,t'} - \Omega^3_{i,t,t'+1} \leq \Delta^3_{i,t,t'} \quad \forall i, t\text{=1,2,..,}T\text{-1 and } t'\text{=}t\text{+1,}t\text{+2,..,}T. \tag{4.26}$$

$$\sum_{i=1}^{N} \sum_{t''=1}^{t-1} \Omega^3_{i,t'',t} \leq 1, \quad t\text{=2,3,..,}T. \tag{4.27}$$

$$\sum_{i'=1}^{N} (\delta^1_{i',t'-1} + \delta^2_{i',t'-1} + \delta^3_{i',t'-1}) \leq N(1 - \Delta^3_{i,t,t'}) \quad \forall i, t\text{=1,2,..,}T\text{-2 and } t'\text{=}t\text{+2,}t\text{+3,..,}T. \tag{4.28}$$

$$\sum_{i'=1}^{N} (\delta^1_{i',t'-1} + \delta^2_{i',t'-1} + \delta^3_{i',t'-1}) \leq N(1 - \Omega^3_{i,t,t'}) \quad \forall i, t\text{=1,2,..,}T\text{-2 and } t'\text{=}t\text{+2,}t\text{+3,..,}T. \tag{4.29}$$

$$X^3_{i,t,t'} \geq \mathcal{E}_d - \frac{C_{t'}}{a_i}(1 - \Delta^3_{i,t,t'}) \quad \forall i, t\text{=1,2,..,}T\text{-1 and } t'\text{=}t\text{+1,}t\text{+2,..,}T. \tag{4.30}$$

$$a_i\, X^3_{i,t,t'} \leq C_{t'}\, \Delta^3_{i,t,t'} \quad \forall i, t\text{=1,2,..,}T\text{-1 and } t'\text{=}t\text{+1,}t\text{+2,..,}T. \tag{4.31}$$

$$s^3_{i,t,t'} \geq \mathcal{E} - C_{t'}(1 - \Omega^3_{i,t,t'}) \quad \forall i, t\text{=1,2,..,}T\text{-1 and } t'\text{=}t,t\text{+1,..,}T. \tag{4.32}$$

$$s^3_{i,t,t'} \leq C_{t'}\, \Omega^3_{i,t,t'} \quad \forall i, t\text{=1,2,..,}T\text{-1 and } t'\text{=}t,t\text{+1,..,}T. \tag{4.33}$$

$$\sum_{t'=t}^{T} s^3_{i,t,t'} \leq ST_i + ((ST_i + 1) \times (1 - \Omega^3_{i,t,t})) \quad \forall i \text{ and } t\text{=1,2,..,}T\text{-1.} \tag{4.34}$$

$$\sum_{t'=t}^{T} s^3_{i,t,t'} \geq ST_i - ((ST_i + 1) \times (1 - \Omega^3_{i,t,t})) \quad \forall i \text{ and } t\text{=1,2,..,}T\text{-1.} \tag{4.35}$$

$$s^3_{i,t,t'-1} \leq C_{t'-1} + ((C_{t'-1} + 1) \times (1 - \Omega^3_{i,t,t'})) \quad \forall i, t\text{=1,2,..,}T\text{-2 and } t'\text{=}t\text{+2,}t\text{+3,..,}T. \tag{4.36}$$

$$s^3_{i,t,t'-1} \geq C_{t'-1} - ((C_{t'-1} + 1) \times (1 - \Omega^3_{i,t,t'})) \quad \forall i, t\text{=1,2,..,}T\text{-2 and } t'\text{=}t\text{+2,}t\text{+3,..,}T. \tag{4.37}$$

$$\Delta^3_{i,t,t} = \Delta^3_{i,t,T+1} = X^3_{i,t,t}$$

$$= X^3_{i,t,T+1} = \Omega^3_{i,t,T+1} = 0 \quad \forall i \text{ and } \forall t. \tag{4.38}$$

$$\delta^3_{i,T} = \Omega^3_{i,T,T} = 0 \quad \forall i. \tag{4.39}$$

/* Constraints (4.40)–(4.58) represent a possible setup crossover across a number of periods with the production starting in period $t' + 1$, when the setup ends at the end-of-period t' */

$$\Omega^4_{i,t,t} \leq \delta^3_{i,t} \qquad \forall i \text{ and } t=1,2,..,T\text{-}2. \tag{4.40}$$

$$\Omega^4_{i,t,t+1} = \Omega^4_{i,t,t} \qquad \forall i \text{ and } t=1,2,..,T\text{-}2. \tag{4.41}$$

$$\Omega^4_{i,t,t'} \geq \Omega^4_{i,t,t'+1} \qquad \forall i,\ t=1,2,..,T\text{-}2 \text{ and } t'=t,t+1,..,T\text{-}1. \tag{4.42}$$

$$\Omega^4_{i,t,t'} + \Delta^4_{i,t,t'} \geq \Delta^4_{i,t,t'+1} \qquad \forall i,\ t=1,2,..,T\text{-}2 \text{ and } t'=t+1,t+2,..,T. \tag{4.43}$$

$$\Delta^4_{i,t,t'} \leq \Omega^4_{i,t,t} \qquad \forall i,\ t=1,2,..,T\text{-}2 \text{ and } t'=t,t+1,..,T. \tag{4.44}$$

$$\Omega^4_{i,t,t'} + \Delta^4_{i,t,t'} \leq 1 \qquad \forall i,\ t=1,2,..,T\text{-}2 \text{ and } t'=t+1,t+2,..,T\text{-}1. \tag{4.45}$$

$$\Omega^4_{i,t,t'} - \Omega^4_{i,t,t'+1} \leq \Delta^4_{i,t,t'+1} \qquad \forall i,\ t=1,2,..,T\text{-}2 \text{ and } t'=t+1,t+2,..,T\text{-}1. \tag{4.46}$$

$$\sum_{i=1}^{N} \sum_{t''=1}^{t-1} \Omega^4_{i,t'',t} \leq 1, \qquad t=2,3,..,T\text{-}1. \tag{4.47}$$

$$\sum_{i'=1}^{N} (\delta^1_{i',t'-1} + \delta^2_{i',t'-1} + \delta^3_{i',t'-1}) \leq N(1 - \Delta^4_{i,t,t'}) \qquad \forall i,\ t=1,2,..,T\text{-}2 \text{ and } t'=t+2,t+3,..,T. \tag{4.48}$$

$$\sum_{i'=1}^{N} (\delta^1_{i',t'-1} + \delta^2_{i',t'-1} + \delta^3_{i',t'-1}) \leq N(1 - \Omega^4_{i,t,t'}) \qquad \forall i,\ t=1,2,..,T\text{-}2 \text{ and } t'=t+2,t+3,..,T. \tag{4.49}$$

$$X^4_{i,t,t'} \geq \mathcal{E}_d - \frac{C_{t'}}{a_i}(1 - \Delta^4_{i,t,t'}) \qquad \forall i,\ t=1,2,..,T\text{-}2 \text{ and } t'=t+2,t+3,..,T. \tag{4.50}$$

$$a_i\, X^4_{i,t,t'} \leq C_{t'}\, \Delta^4_{i,t,t'} \qquad \forall i,\ t=1,2,..,T\text{-}2 \text{ and } t'=t+2,t+3,..,T. \tag{4.51}$$

$$s^4_{i,t,t} \geq \mathcal{E} - C_t(1 - \Omega^4_{i,t,t}) \qquad \forall i \text{ and } t=1,2,..,T\text{-}2. \tag{4.52}$$

$$s^4_{i,t,t} \leq C_t\, \Omega^4_{i,t,t} \qquad \forall i \text{ and } t=1,2,..,T\text{-}2\,. \tag{4.53}$$

$$\sum_{t'=t}^{T-1} s^4_{i,t,t'} \leq ST_i + ((ST_i + 1) \times (1 - \Omega^4_{i,t,t})) \qquad \forall i \text{ and } t=1,2,..,T\text{-}2. \tag{4.54}$$

$$\sum_{t'=t}^{T-1} s^4_{i,t,t'} \geq ST_i - ((ST_i + 1) \times (1 - \Omega^4_{i,t,t})) \qquad \forall i \text{ and } t=1,2,..,T\text{-}2. \tag{4.55}$$

$$s^4_{i,t,t'} = C_{t'}\, \Omega^4_{i,t,t'} \qquad \forall i,\ t=1,2,..,T\text{-}2 \text{ and } t'=t+1,t+2,..,T\text{-}1. \tag{4.56}$$

$$\Delta^4_{i,t,t} = \Delta^4_{i,t,t+1} = \Delta^4_{i,t,T+1} = X^4_{i,t,T+1}$$

$$= X^4_{i,t,t} = X^4_{i,t,t+1} = 0 \qquad \forall i \text{ and } \forall t. \tag{4.57}$$

$$s^4_{i,t,T} = \Omega^4_{i,t,T} = 0 \qquad \forall i \text{ and } \forall t. \tag{4.58}$$

/* Constraints (4.59) and (4.60) indicate that either $\Omega^3_{i,t,t}$ or $\Omega^4_{i,t,t}$ can only exist in period t when $\delta^3_{i,t} = 1$ */

$$\Omega^3_{i,t,t} + \Omega^4_{i,t,t} = \delta^3_{i,t} \qquad \forall i \text{ and } t\text{=1,2,..,}T\text{-1}. \tag{4.59}$$

$$\Omega^3_{i,t,t'} + \Omega^4_{i,t,t'} \leq \delta^3_{i,t} \qquad \forall i, t\text{=1,2,..,}T\text{-1 and } t'\text{=}t,t\text{+1,..,}T. \tag{4.60}$$

/* Constraints (4.61) and (4.62) represent the condition that only one type of production carryover can take place for at most one product in a period */

$$\textstyle\sum_{i=1}^{N}(\Delta^1_{i,1,2} + \Delta^2_{i,1,2} + \Delta^3_{i,1,2}) \leq 1. \tag{4.61}$$

$$\textstyle\sum_{i=1}^{N}\sum_{t''=1}^{t-1}(\Delta^1_{i,t'',t} + \Delta^2_{i,t'',t} + \Delta^3_{i,t'',t})$$

$$+\textstyle\sum_{i=1}^{N}\sum_{t''=1}^{t-2}\Delta^4_{i,t'',t} \leq 1, \qquad t\text{=3,4,..,}T. \tag{4.62}$$

/* Constraints (4.63)–(4.65) show that if the setup of at least 1 product is initiated in period t (indicated by $delta_t$=1), then it means that no carryover of either setup or production of any product, with its setup having been initiated in any period up to $t-1$, can be carried over to period $t+1$ and later. This is so because if any product's setup is initiated in period t, then either the production carryover or setup carryover with the setup initiated up to period $t-1$ and carried over to period $t+1$ and later, is not feasible */

$$\textstyle\sum_{i=1}^{N}\sum_{t''=1}^{t-1}\sum_{t'''=t+1}^{T}(\Delta^1_{i,t'',t'''} + \Delta^2_{i,t'',t'''}$$

$$+\Delta^3_{i,t'',t'''} + \Delta^4_{i,t'',t'''} + \Omega^3_{i,t'',t'''}$$

$$+\Omega^4_{i,t'',t'''}) \leq N \times T^2(1 - delta_t), \qquad t\text{=2,3,..,}T\text{-1}. \tag{4.63}$$

$$\textstyle\sum_{i=1}^{N}(\delta^1_{i,t} + \delta^2_{i,t} + \delta^3_{i,t}) \leq N \times delta_t, \qquad t\text{=2,3,..,}T\text{-1}. \tag{4.64}$$

$$delta_t \leq \textstyle\sum_{i=1}^{N}(\delta^1_{i,t} + \delta^2_{i,t} + \delta^3_{i,t}), \qquad t\text{=2,3,..,}T\text{-1}. \tag{4.65}$$

/* Constraints (4.66) and (4.67) indicate that a production carryover for any product i can take place in period $t+2 \leq t' \leq T$, only if the sum of possible production times and setup times of all products in period $t'-1$ is equal to the capacity of period $t'-1$ */

$$\sum_{i=1}^{N}(a_i\,X_{i,t,t'-1}^{1} + a_i\,X_{i,t,t'-1}^{2} + a_i\,X_{i,t,t'-1}^{3}$$

$$+a_i\,X_{i,t,t'-1}^{4} + s_{i,t,t'-1}^{3} + s_{i,t,t'-1}^{4}) \leq C_{t'-1}+$$

$$((C_{t'-1}+1)\times(1-\sum_{i=1}^{N}(\Delta_{i,t,t'}^{1} + \Delta_{i,t,t'}^{2}$$

$$+\Delta_{i,t,t'}^{3} + \Delta_{i,t,t'}^{4}))), \qquad t{=}1,2,..,T\text{-}2 \text{ and } t'{=}t{+}2,t{+}3,..,T. \tag{4.66}$$

$$\sum_{i=1}^{N}(a_i\,X_{i,t,t'-1}^{1} + a_i\,X_{i,t,t'-1}^{2} + a_i\,X_{i,t,t'-1}^{3}$$

$$+a_i\,X_{i,t,t'-1}^{4} + s_{i,t,t'-1}^{3} + s_{i,t,t'-1}^{4}) \geq C_{t'-1}-$$

$$((C_{t'-1}+1)\times(1-\sum_{i=1}^{N}(\Delta_{i,t,t'}^{1} + \Delta_{i,t,t'}^{2}$$

$$+\Delta_{i,t,t'}^{3} + \Delta_{i,t,t'}^{4}))), \qquad t{=}1,2,..,T\text{-}2 \text{ and } t'{=}t{+}2,t{+}3,..,T. \tag{4.67}$$

/* Constraints (4.68)–(4.70) represent the feasibility of production and setup with respect to capacity availability */

$$\sum_{i=1}^{N}(a_i\,X_{i,1,1}^{1}+s_{i,1}^{1}+s_{i,1}^{2}+s_{i,1,1}^{3}+s_{i,1,1}^{4}) \leq C_1. \tag{4.68}$$

$$\sum_{i=1}^{N}(s_{i,t}^{1} + s_{i,t}^{2} + \sum_{t''=1}^{t} s_{i,t'',t}^{3} + \sum_{t''=1}^{t} s_{i,t'',t}^{4})$$

$$+\sum_{i=1}^{N}\sum_{t''=1}^{t} a_i\,X_{i,t'',t}^{1}$$

$$+\sum_{i=1}^{N}\sum_{t''=1}^{t-1}(a_i\,X_{i,t'',t}^{2} + a_i\,X_{i,t'',t}^{3}$$

$$+a_i\,X_{i,t'',t}^{4}) \leq C_t, \qquad t{=}2,3,..,T\text{-}1. \tag{4.69}$$

$$\sum_{i=1}^{N} s_{i,T}^{1} + \sum_{i=1}^{N}\sum_{t''=1}^{T-1} s_{i,t'',T}^{3}$$

$$+\sum_{i=1}^{N}\sum_{t''=1}^{T} a_i\,X_{i,t'',T}^{1}$$

$$+\sum_{i=1}^{N}\sum_{t''=1}^{T-1}(a_i\,X_{i,t'',T}^{2} + a_i\,X_{i,t'',T}^{3}$$

$$+a_i\,X_{i,t'',T}^{4}) \leq C_T. \tag{4.70}$$

/* Constraints (4.71)–(4.72) represent the inventory balance constraints; constraint (4.73) assumes that there is neither inventory nor backorder of any product at the end of the planning horizon */

$$I_{i,1} - B_{i,1} = X_{i,1,1}^{1} - d_{i,1} \qquad \forall i. \tag{4.71}$$

$$I_{i,t} - B_{i,t} = \sum_{t'''=1}^{t} \sum_{t''=1}^{t'''} X^1_{i,t'',t'''}$$

$$+ \sum_{t'''=2}^{t} \sum_{t''=1}^{t'''-1} (X^2_{i,t'',t'''} + X^3_{i,t'',t'''}$$

$$+ X^4_{i,t'',t'''}) - \sum_{t'''=1}^{t} d_{i,t'''} \qquad \forall i \text{ and } t=2,3,..,T. \qquad (4.72)$$

$$I_{i,T} = B_{i,T} = 0 \qquad \forall i. \qquad (4.73)$$

$\delta^1_{i,t}$, $\delta^2_{i,t}$, $\delta^3_{i,t}$, $\Delta^1_{i,t,t'}$, $\Delta^2_{i,t,t'}$, $\Delta^3_{i,t,t'}$, $\Delta^4_{i,t,t'}$, $\Omega^3_{i,t,t'}$, $\Omega^4_{i,t,t'}$ are binary for all i, t and t' (with $t \leq t' \leq T$ or $t+1 \leq t' \leq T$, as appropriate to the respective binary variable), and all other variables are ≥ 0.
Note: $\Omega^4_{i,T-1,T-1} = \Omega^4_{i,T-1,T} = \Delta^4_{i,T-1,T} = \Delta^4_{i,T-1,T+1} = \delta^1_{i,0} = \delta^2_{i,0} = \delta^3_{i,0} = 0 \; \forall i$.

Also in Eq. (4.71), $B_{i,T} = 0 \; \forall i$ exists when backorders are not supposed to be present at the end of the planning horizon (i.e. here the demand for all products should be satisfied within period T). In contrary, the equation must be removed when demand for all products need not be satisfied within period T. It also implies that backorders are allowed during the planning period, but no backorder is allowed beyond the planning horizon.

The objective function shown in Eq. (4.1) is the minimization of setup cost, holding cost and backorder cost of all products across all time periods. Constraint (4.2) indicates that there can be either an end-of-period setup or a setup crossover for at most one product in every period t. Constraint (4.3) indicates that for every product i, there can be at most one of the setup-types in each period t. Constraints (4.4) and (4.11) indicate that when there is a setup of product i in time period t, it means that a production is carried out in period t' corresponding to the setup in period t. Constraints (4.5) and (4.12) ensure that a production carryover in period $t'+1$ is not possible without a corresponding production carryover in period t'. Constraints (4.6), (4.13), (4.28) and (4.48) indicate that a production can be carried over to period t' due to the setup in period t without any interruption, only if there is no intermittent setup of any other product. Constraints (4.7), (4.8), (4.14), (4.15), (4.30), (4.31), (4.50) and (4.51) are the production constraints. Here, constraints (4.7), (4.14), (4.30) and (4.50) address the condition that a production exists corresponding to the indicator variable (i.e. when the indicator variable takes the value 1, at least a small $\mathcal{E}_d$ unit of the corresponding product should be produced). Constraints (4.9), (4.16), (4.34), (4.35), (4.54) and (4.55) ensure that when the setup of product i is initiated in period t (i.e. when $\delta^1_{i,t}$ or $\delta^2_{i,t}$ or $\Omega^3_{i,t,t}$ or $\Omega^4_{i,t,t}$ exists), the total setup time ST_i required by product i for setup is consumed. Constraints (4.10), (4.17), (4.32), (4.33), (4.36), (4.37), (4.52), (4.53) and (4.56) represent the conditions for a setup to be present in a period. Constraints (4.20)–(4.22), (4.40)–(4.42) and (4.59)–(4.60) indicate the conditions for a setup to cross over across periods. Constraints (4.23)–(4.26) and (4.43)–(4.46) indicate the conditions for a production to occur in a period due to a setup crossover. Constraints (4.27) and (4.47) show the condition that a setup can be present in period t, due to a setup crossover initiated from at most one of the preceding periods and can occur for at most one product. Constraints (4.29) and (4.49) indicate that when a setup is carried over across periods, no setup of any product should exist in the intermittent periods. Constraints (4.61) and (4.62) represent the condition that only one type of production carryover can take place for at most one product in a period. Constraints (4.63)–(4.65) show that if the setup of at least 1 product is initiated in period t (indicated by $delta_t = 1$), then it means that no carryover of either setup or production of any product, with its setup having been initiated in any period up to $t-1$, can be carried over to period $t+1$ and later. This is so because if any product's setup is initiated in period t, then either the production carryover or setup carryover with the setup initiated up to period $t-1$ and carried over to period $t+1$ and later is not feasible. Constraints (4.66)–(4.67) state that the production carryover of a product between adjacent periods should not be interrupted with idle time, i.e., the constraints indicate that the production carryover of product i can take place in period t' due to the setup of product i in period t, only if the sum of the possible setup times and production times of all products is equal to the capacity of the previous period t'-1. Constraints (4.68)–(4.70) denote the capacity constraints, with the consideration of possible setup times and production times of all products. Constraints (4.71)–(4.72) represent the inventory balance constraints. Constraint (4.73) assumes that there is neither inventory nor backorder of any product at the end of the planning horizon. Constraints (4.18), (4.19), (4.38), (4.39), (4.57) and (4.58) are the boundary conditions for the mathematical model. It is to be noted that the parameters $\mathcal{E}_d$ and $\mathcal{E}$ are set to a small positive real value.

4.3.4 Method of Tracking Setups in MM2:CLSP-PCSC

A brief explanation on the principle of tracking the setups in MM2:CLSP-PCSC especially for multi-period setups (within a time period/across time periods), the corresponding setup times and production periods is given in this section.

The variable $\delta^3_{i,t}$ is used to track the multi-period setup, and the variables $\Omega^3_{i,t,t'}$ and $\Omega^4_{i,t,t'}$ are linked with $\delta^3_{i,t}$. $\Omega^3_{i,t,t'}$ tracks the setup which starts in period t and runs across one or more periods (t') and gets completed before, and not at the end of a future period. $\Omega^4_{i,t,t'}$ tracks the setup which starts in period t and runs across one or more periods (t') and gets completed at the end of a future period. In variables $\Omega^3_{i,t,t'}$ and $\Omega^4_{i,t,t'}$, t refers to the period where the setup starts and t' refers to a future period where the setup time is present.

In MM2:CLSP-PCSC, the setup times corresponding to the setup are tracked using variables $s^3_{i,t,t'}$ and $s^4_{i,t,t'}$, where t represents the period in which the setup starts and t' represents a future period in which the setup time is present.

The variables $\Delta^3_{i,t,t'}$ and $\Delta^4_{i,t,t'}$ (corresponding to $\Omega^3_{i,t,t'}$ and $\Omega^4_{i,t,t'}$) track the production across periods due to the multi-period setups, where t represents the period in which the setup starts and t' represents a future period in which the production may occur.

4.4 Special Cases of CLSP-PCSC with Respect to MM2:CLSP-PCSC

Two special cases of CLSP-PCSC with respect to MM2:CLSP-PCSC are presented in this section. They are: (i) setup cost of a product calculated with respect to the period of its setup completion and (ii) setup cost and holding cost of a product are time independent.

4.4.1 Setup Cost of a Product Calculated with Respect to the Period of Its Setup Completion

When the cost of setup for a product is computed with respect to the period where its setup ends, the objective function of MM2:CLSP-PCSC is modified and some constraints are added to the model, apart from the previous set of constraints.

Variable	Description
$\psi^1_{i,t}$	An indicator (binary) variable that takes value 1: it corresponds to the setup of product i in period t when $\delta^1_{i,t} = 1$; 0 otherwise.
$\psi^2_{i,t}$	An indicator (binary) variable that takes value 1: it corresponds to the setup of product i at the end of period t when $\delta^2_{i,t} = 1$; 0 otherwise.
$\psi^3_{i,t,t'}$	An indicator (binary) variable that takes value 1 in the period when the setup of product i ends, when the setup of product i is initiated in period t and is present in period t' ($t \leq t' \leq T$), with the setup of product i ending in a period later that period t, but not exactly at the end of that period. *Note:* the variable can take the value of 1 provided $\Omega^3_{i,t,t'} = 1$; 0 otherwise.
$\psi^4_{i,t,t'}$	An indicator (binary) variable that takes value 1 in the period when the setup of product i ends, when the setup of product i is initiated in period t and is present in period t' ($t \leq t' \leq T-1$), with the setup of product i ending in a period later than period t, and the setup getting completed exactly at the end of that period. *Note:* the variable can take the value of 1 provided $\Omega^4_{i,t,t'} = 1$; 0 otherwise.

The objective function of Mathematical Model 2 is:

$$\text{Min} \quad Z = \sum_{i=1}^{N}\sum_{t=1}^{T} SC_{i,t}\,\psi^1_{i,t} + \sum_{i=1}^{N}\sum_{t=1}^{T-1} SC_{i,t}\,\psi^2_{i,t} + \sum_{i=1}^{N}\sum_{t=1}^{T-1}\sum_{t'=t+1}^{T} SC'_{i,t'}\,\psi^3_{i,t,t'} + \tag{4.74}$$
$$\sum_{i=1}^{N}\sum_{t=1}^{T-2}\sum_{t'=t+1}^{T-1} SC'_{i,t'}\,\psi^4_{i,t,t'} + \sum_{i=1}^{N}\sum_{t=1}^{T} h_{i,t}\,I_{i,t} + \sum_{i=1}^{N}\sum_{t=1}^{T} b_i\,B_{i,t}$$

Subject to constraints (4.2)–(4.73).
Where, $SC'_{i,t'}$—setup cost for product i with its setup getting completed in period t'; this cost is incurred only once as a fixed cost computed with respect to the period of its setup completion, *Note:* $SC_{i,t} = SC'_{i,t'}$ with respect to $\psi^1_{i,t}$ and $\psi^2_{i,t}$.
In addition to this, constraints (4.75)–(4.86) are added to the model. They are as follows.

$$\psi^1_{i,t} = \delta^1_{i,t} \qquad \forall i \text{ and } \forall t. \tag{4.75}$$

$$\psi^2_{i,t} = \delta^2_{i,t} \qquad \forall i \text{ and } t=1,2,..,T\text{-}1. \tag{4.76}$$

$$\psi^3_{i,t,t'} \leq \Omega^3_{i,t,t'} \qquad \forall i, t=1,2,..,T\text{-}1 \text{ and } t'=t,t+1,..,T. \tag{4.77}$$

$$\Omega^3_{i,t,t'} - \Omega^3_{i,t,t'+1} \leq \psi^3_{i,t,t'} \qquad \forall i, t=1,2,..,T\text{-}1 \text{ and } t'=t+1,t+2,..,T. \tag{4.78}$$

$$\Omega^3_{i,t,t'} + \Omega^3_{i,t,t'+1} \leq 2 - \psi^3_{i,t,t'} \qquad \forall i, t=1,2,..,T\text{-}1 \text{ and } t'=t+1,t+2,..,T. \tag{4.79}$$

$$\sum_{t'=t}^{T} \psi^3_{i,t,t'} = \Omega^3_{i,t,t} \qquad \forall i \text{ and } t=1,2,..,T\text{-}1. \tag{4.80}$$

$$\psi^3_{i,t,t} = \psi^3_{i,t,T+1} = 0 \qquad \forall i \text{ and } \forall t. \tag{4.81}$$

$$\psi^4_{i,t,t'} \leq \Omega^4_{i,t,t'} \qquad \forall i, t=1,2,..,T\text{-}2 \text{ and } t'=t,t+1,..,T\text{-}1. \tag{4.82}$$

$$\Omega^4_{i,t,t'} - \Omega^4_{i,t,t'+1} \leq \psi^4_{i,t,t'} \qquad \forall i, t=1,2,..,T\text{-}2 \text{ and } t'=t+1,t+2,..,T\text{-}1. \tag{4.83}$$

$$\Omega^4_{i,t,t'} + \Omega^4_{i,t,t'+1} \leq 2 - \psi^4_{i,t,t'} \qquad \forall i, t=1,2,..,T\text{-}2 \text{ and } t'=t+1,t+2,..,T\text{-}1. \tag{4.84}$$

$$\sum_{t'=t}^{T-1} \psi^4_{i,t,t'} = \Omega^4_{i,t,t} \qquad \forall i \text{ and } t=1,2,..,T\text{-}2. \tag{4.85}$$

$$\psi^4_{i,t,t} = \psi^4_{i,t,T} = 0 \qquad \forall i \text{ and } \forall t. \tag{4.86}$$

4.4.2 Setup Cost and Holding Cost of a Product Being Time Independent

When the cost for setup is product dependent and time independent, the objective function for MM2:CLSP-PCSC is modified as follows:

$$\text{Min } Z = \sum_{i=1}^{N}\sum_{t=1}^{T} SC'_i\, \delta^1_{i,t} + \sum_{i=1}^{N}\sum_{t=1}^{T-1} SC'_i\, \delta^2_{i,t} + \sum_{i=1}^{N}\sum_{t=1}^{T-1} SC'_i\, \delta^3_{i,t} + \sum_{i=1}^{N}\sum_{t=1}^{T} h'_i\, I_{i,t} + \sum_{i=1}^{N}\sum_{t=1}^{T} b_i\, B_{i,t} \tag{4.87}$$

Subject to constraints (4.2)–(4.73).
Where, SC'_i—setup cost for product i, and likewise for h'_i.

4.5 Numerical Illustrations and Discussion with Respect to MM2:CLSP-PCSC

The same sample problem instance shown in Table 4.1 is considered to illustrate the generalized version of the model MM2:CLSP-PCSC and its special case, namely 'Setup cost of a product calculated with respect to the period of its setup completion'. Values for the time-independent setup costs for the five products are $SC'_i = \{50, 10, 30, 20, 40\},\ \ i = 1, 2, 3, 4, 5$, and time-independent holding costs for the five products are $h'_i = \{2, 3, 1, 1, 2\},\ \ i = 1, 2, 3, 4, 5$. For the sake of illustration the proposed model assumes equal capacity across different periods. However the proposed model allows for different capacities across different time periods due to the use of C_t, without loss of generality. These values along with the values $d_{i,t}$, ST_i, b_i and a_i provided in Table 3.1 are used to illustrate a special case, namely 'Setup cost and holding cost of a product are time independent'.

Table 4.1: A sample problem instance

Product	Setup time (ST_i): (time units)	Backorder cost (b_i): (mu/period/unit product backordered)	Number of time units required for producing one unit of product i (a_i): (time units/unit of product i)
1	10	20	1
2	220	20	1
3	200	40	1
4	10	30	1
5	100	30	1

Product \ Period	1	2	3	4	5	6	7	8	9	10	11	12	13	14	15	16
Demand ($d_{i,t}$)																
1	60	0	0	0	0	0	0	0	0	0	0	0	0	0	0	0
2	0	0	0	90	20	0	0	0	0	0	0	0	0	0	0	0
3	0	0	0	0	0	0	45	0	0	0	0	0	0	0	0	100
4	0	0	0	0	0	0	0	80	70	60	0	0	50	0	0	0
5	0	0	0	0	0	0	0	0	0	0	0	100	10	0	0	0
Setup cost ($SC_{i,t}$): (mu/setup)																
1	20	30	10	30	50	20	40	20	40	30	20	10	20	40	30	20
2	40	20	40	30	20	10	10	20	10	30	40	20	10	40	60	50
3	30	50	60	40	20	40	30	60	40	20	10	30	40	20	10	20
4	20	40	30	20	10	30	60	30	60	20	50	20	10	30	60	20
5	60	30	20	20	30	10	30	30	20	60	40	40	50	20	10	30
Holding cost ($h_{i,t}$): (mu/period/unit product carried over)																
1	1	1	1	1	1	2	2	2	2	3	3	3	3	3	3	3
2	2	2	2	2	1	1	1	1	1	2	2	4	4	4	4	4
3	4	4	4	4	3	3	3	3	2	2	2	2	1	1	1	1
4	3	3	3	2	2	2	1	1	3	1	1	2	2	2	3	3
5	4	4	4	1	1	1	3	3	3	4	4	4	2	2	2	2
Capacity (C_t): time units	100	100	100	100	100	100	100	100	100	100	100	100	100	100	100	100

4.5.1 Setup Cost of a Product Calculated with Respect to the Period of Its Setup Initiation

When the generalized Mathematical Model 2 is executed for the same problem instance given in Table 4.1, a solution (see Table 4.2) with its corresponding Gantt chart shown in Fig. 4.2 is obtained, with the value of the objective function equal to 200 mu (monetary units). Product 4 is produced with two setups (see periods 8, 9 and 10) with a break in production (maintaining the feasibility condition with respect to production occurring immediately after setup and no break in production carryover). This is so because the setup cost of product 4 is smaller than its holding cost up to that period. For the same problem instance given in Table 4.1, the setup cost of product 4 is increased from 60 mu to 120 mu in period 9, and is increased from 20 mu to 120 mu in period 10. Now it is observed that product 4 is continuously produced in periods 8, 9 and 10 with a single setup (see Fig. 4.3). This is so because the setup cost of product 4 is greater that its holding cost. The objective function value obtained in this case (when MM2:CLSP-PCSC is executed) is equal to $Z = 270$ mu. Through these observations it is understood that both the mathematical models address situations where an uninterrupted production has to take place (see Fig. 4.3), and if there is any discontinuity or break in production across periods (due to a lower setup cost compared to the holding cost) a new setup can be done for the same product for any subsequent production (see Fig. 4.2). Hence, both the mathematical models can address scenarios specific to process industries.

4.5.2 Setup Cost of a Product Calculated with Respect to the Period of Its Setup Completion

The case where the setup cost is calculated in period t' (period when the setup ends) is illustrated with the help of the same numerical example shown in Table 4.1. Here the products are assumed to havc time-dependent and product-dependent setup costs ($SC'_{i,t'}$) and holding costs ($h_{i,t}$), whose values are given in Table 4.1. The Gantt chart corresponding the solution obtained is shown in Fig. 4.4. The value of the objective function obtained is 190 mu.

4.5.3 Setup Cost and Holding Cost of a Product Being Time Independent

The time-independent cost structure (i.e. SC'_i and h'_i whose values are provided in this section) is considered along with the demand and backorder cost data (i.e. $d_{i,t}$ and b_i) shown in Table 4.1 to illustrate this special case. The Gantt chart corresponding the solution obtained is shown in Fig. 4.5. The value of the objective function obtained is 220 mu.

Table 4.2: Solution generated by the proposed mathematical model MM2:CLSP-PCSC (corresponding terms in MM2:CLSP-PCSC are used here) for the data given in Table 3.1 with the corresponding Gantt chart provided in Fig. 4.2

$t = 1$	$t = 2$	$t = 3$	$t = 4$	$t = 5$	$t = 6$	$t = 7$	$t = 8$
$\delta^1_{1,1} = 1$ $\delta^1_{1,1} = 1$ $\Delta^1_{1,1,1} = 1$ $X^1_{1,1,1} = 60$ $s^1_{1,1} = 10$ $s^3_{2,1,1} = 10$ $\Omega^3_{2,1,1} = 1$	$s^3_{2,1,2} = 100$ $\Omega^3_{2,1,2} = 1$	$s^3_{2,1,3} = 100$ $\Omega^3_{2,1,3} = 1$	$\Delta^3_{2,1,4} = 1$ $X^3_{2,1,4} = 90$ $s^3_{2,1,4} = 10$ $\Omega^3_{2,1,4} = 1$	$\delta^3_{3,5} = 1$ $\Delta^3_{2,1,5} = 1$ $X^3_{2,1,5} = 20$ $s^3_{3,5,5} = 80$ $\Omega^3_{3,5,5} = 1$	$s^3_{3,5,6} = 100$ $\Omega^3_{3,5,6} = 1$	$\Delta^3_{3,5,7} = 1$ $X^3_{3,5,7} = 45$ $s^3_{3,5,7} = 20$ $\Omega^3_{3,5,7} = 1$	$\delta^1_{4,8} = 1$ $\Delta^1_{4,8,8} = 1$ $X^1_{4,8,8} = 80$ $s^1_{4,8} = 10$
$t = 9$	$t = 10$	$t = 11$	$t = 12$	$t = 13$	$t = 14$	$t = 15$	$t = 16$
$\Delta^1_{4,8,9} = 1$ $X^1_{4,8,9} = 70$	$\delta^1_{4,10} = 1$ $\Delta^1_{4,10,10} = 1$ $X^1_{4,10,10} = 60$ $s^1_{4,10} = 10$	$\delta^2_{5,11} = 1$ $s^2_{5,11} = 100$	$\Delta^2_{5,11,12} = 1$ $X^2_{5,11,12} = 100$	$\delta^1_{4,13} = 1$ $\Delta^1_{4,13,13} = 1$ $\Delta^2_{5,11,13} = 1$ $X^1_{4,13,13} = 50$ $X^2_{5,11,13} = 10$ $s^1_{4,13} = 10$	$\delta^3_{3,14} = 1$ $s^4_{3,14,14} = 100$ $\Omega^4_{3,14,14} = 1$	$s^4_{3,14,15} = 100$ $\Omega^4_{3,14,15} = 1$	$\Delta^4_{3,14,16} = 1$ $X^4_{3,14,16} = 100$

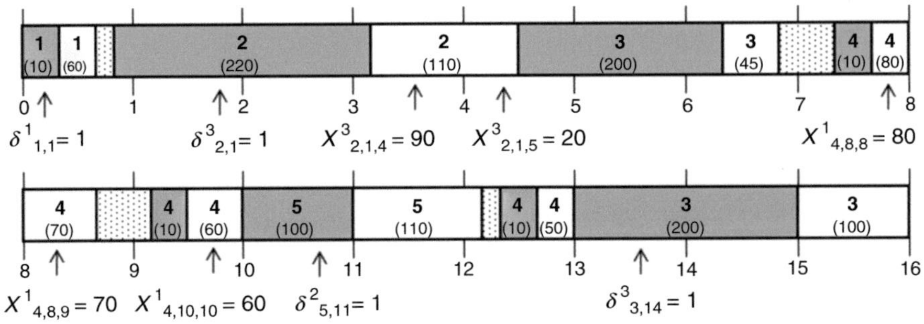

Figure 4.2: Gantt chart for the solution obtained by the proposed model MM2:CLSP-PCSC (for the data given in Table 4.1 and the solution provided in Table 4.2) when the setup cost and holding cost of a product are time dependent, and the setup cost is calculated with respect to the period of its setup initiation; $Z = 200$ mu

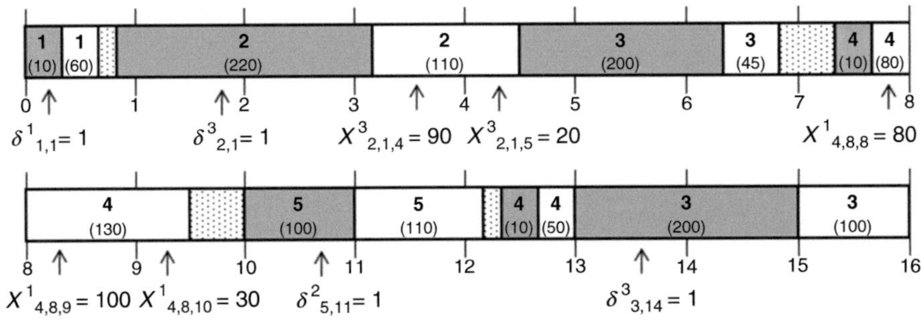

Figure 4.3: Gantt chart for the solution obtained by the proposed model MM2:CLSP-PCSC when the setup cost and holding cost of a product are time dependent (for the data given in Table 4.1), and the setup cost is calculated with respect to the period of its setup initiation. Here the setup cost for product 4 is increased from 60 mu to 120 mu in period 9, and from 20 mu to 120 mu in period 10; $Z = 270$ mu

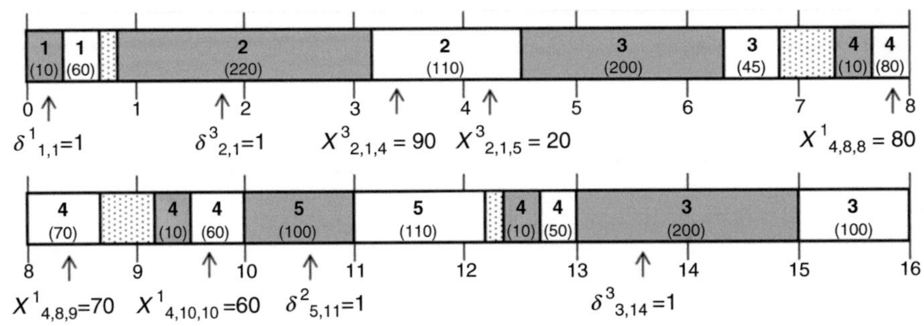

Figure 4.4: Gantt chart obtained by the proposed model MM2:CLSP-PCSC (for the data given in Table 4.1) when the setup cost and holding cost of a product are time dependent, and the setup cost is calculated with respect to the period of its setup completion; $Z = 190$ mu

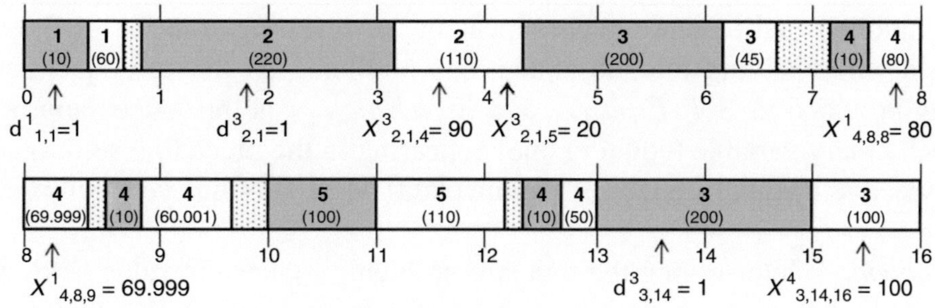

Figure 4.5: Gantt chart for the solution obtained by the proposed model MM2:CLSP-PCSC (for the data given in Sect. 4.5 and Table 4.1) when the setup costs and holding costs are time independent; $Z = 220$ mu

The result for all the above cases is similar to the one obtained by MM1:CLSP-PCSC.

4.6 Proposed Heuristic for CLSP-PCSC with Respect to MM2:CLSP-PCSC

A heuristic based on MM2:CLSP-PCSC is proposed in this section. The heuristic presented for MM1:CLSP-PCSC assumes identical C_t's across all time periods, and the shift of setup/production is restricted to one period ahead of or after the current time period. However, when non-identical C_t's are present across time periods, allowing shift of setup/production for more periods ahead of or after the current time period, the extension of the heuristic based on MM1:CLSP-PCSC becomes tedious. In such cases, with these considerations, we need to use the heuristic based on MM2:CLSP-PCSC. The proposed heuristic with respect to MM2:CLSP-PCSC uses the concept of Relax-and-Fix, with limited similarity to related works of the authors mentioned in Sect. 3.6; however, it has its own novelty in terms of determining the size of the time window and solving each sub-problem. The proposed CLSP-PCSC heuristic is illustrated with the help of Fig. 3.8 (in Chap. 3) and the steps are briefly explained using a pseudocode. Here $MoveWindow$ refers to the window up to which all the binary variables are set to be binary integers, starting from period 1, i.e., all the binary variables $\in \{0, 1\}$ (B_{MIP} refers to binary integers set as binary in $MoveWindow$), and $FinWindow$ is the window in which all the binary integer variables are relaxed as continuous variables within the interval [0,1], up to period T (LP_R refers to relaxation of binary variables in all periods after $MoveWindow$). $WinSize$ refers to the size of the window (expressed in time periods) which is assigned to $MoveWindow$. The size of a time window for every problem instance (see Eq. (4.88)) is expressed as the ceiling value or the smallest integer value greater than or equal to the $\lceil$ maximum setup time (among all products expressed in time units) divided by the minimum capacity (among all periods expressed in time units) $\rceil$ (rounded off to the higher integer if necessary)+1 (expressed in time periods).

$$WinSize \text{ (expressed in time periods)} = \left\lceil \frac{max_i\{ST_i\}\text{(expressed in time units)}}{min_t\{C_t\}\text{(expressed in time units)}} \right\rceil + 1 \quad (4.88)$$

The symbol τ (initially set to zero) is used to refer to the time period up to which the binary variables are set before solving a sub-problem. Once a sub-problem is solved, the binary variables (solution) equal to one which appear in the resulting solution of the sub-problem are considered while solving the subsequent sub-problem. In the subsequent sub-problem, the inequality provided in $SolnEqu_{1,i,t,\tau}$ and $SolnEqu_{2,i,t,\tau}$ is introduced appropriately with respect to each binary variable (equal to one) appearing in the preceding solution (until period τ). These respective inequalities are set (i.e. invoked) while solving every sub-problem.

First Sub-Problem: While solving the first sub-problem, τ takes the value zero, which means that none of the periods are preset with any inequality ($SolnEqu's$) in the first iteration. In the first sub-problem, all binary variables (in $MoveWindow$) $\in \{0, 1\}$ and all the binary variables in $FinWindow$ are relaxed in the interval [0,1], and the model is solved using an MIP solver.

Subsequent Sub-Problem Until $(T - WinSize + 1)$*th Sub-Problem:* In the subsequent sub-problems, τ is incremented to τ+1. Using the resulting solution values from the preceding sub-problem (i.e. the binary variables with the value 1, present up to the τth time period obtained from the previous sub-problem) are considered and each such binary variable leads to the corresponding introduction to one of $SolnEqu_{1,i,t,\tau}$ or $SolnEqu_{2,i,t,\tau}$, as appropriate; such binary variables are now set as $\{0, 1\}$ and not equal to 1, while solving the present sub-problem. The corresponding inequalities are set (i.e. invoked) while solving the current sub-problem due to which, either a left shift or a right shift or *status quo* of a setup/production is allowed to occur, while solving the subsequent sub-problems so as to possibly improve the solution. In MM2:CLSP-PCSC, shift of setup/production is not restricted to one period. $HWinSize$ helps to choose the number of periods up to which the shift of setup/production is allowed to occur. $HWinSize \in 1, 2, \ldots, WinSize$, i.e., any value between 1 and $WinSize$ can be chosen to fix the number of shifts that can be made, by introducing the $SolnEqu's$ corresponding to the binary variable (solution) generated. For example, if the chosen value of $HWinSize$=2, when the value of $WinSize$ is 4 time periods, the inequalities introduced ensure at least one setup initiation of the corresponding product i appropriately in periods $t - 2$, $t - 1$, t, $t + 1$ or $t + 2$ (i.e. if product i is set up in period t, then the shift of setup/production is restricted to two periods ahead of or two periods after the current time period in which the setup in initiated). Therefore, while solving the current sub-problem, the $SolnEqu_{1,i,t,\tau}$ and $SolnEqu_{2,i,t,\tau}$ are introduced corresponding to the generated solution variables (of the previous sub-problem) up to period τ; periods from $\tau + 1$ to $\tau + WinSize$ are considered as the next $MoveWindow$ up to which all the binary variables are considered as belonging to $\{0, 1\}$ right from time period 1; and $FinWindow$ starts from period $\tau + WinSize + 1$ and goes up to period T where all binary variables are relaxed as continuous variables in the interval [0,1]. Note that $SolnEqu_{1,i,t,\tau}$ and $SolnEqu_{2,i,t,\tau}$ are not considered in the $MoveWindow$. Such a sub-problem (with such a setting of inequalities up to period τ, setting of binary variables up to $\tau + WinSize \in \{0, 1\}$ from period 1, and all binary variables relaxed as continuous variables in the interval [0,1] from period $(\tau + WinSize + 1)$ up to period T) is solved using an MIP solver. This procedure continues until all sub-problems are solved up to the complete time horizon. It should be noted that binary variables (solution) are present as either 0 or 1 (without being relaxed) in the final solution. The number of sub-problems to be solved for every instance in the proposed heuristic is equal to $T - WinSize + 1$.

Note: In the heuristic with respect to MM1:CLSP-PCSC only one left shift and right shift of setup/production ahead of or after the current time period is allowed. However, since the extension of heuristic of MM1:CLSP-PCSC is tedious when the capacity is unequal across periods, the heuristic based on MM2:CLSP-PCSC is proposed. Here $HWinSize$ refers to number of left shifts and right shifts that can be allowed in the heuristic, where $WinSize$ amounts to the maximum number of shifts that can be allowed.

Pseudocode for the Proposed Heuristic Based on MM2:CLSP-PCSC

1: **function**
Heuristic ($MoveWindow, FinWindow, WinSize, HWinSize,$
$\tau, t, SolnEqu_{1,i,t,\tau}, SolnEqu_{2,i,t,\tau}$)

2: **Compute**

$$WinSize\text{(expressed in time units)} = \left\lceil \frac{max_i\{ST_i\}\text{(expressed in time units)}}{min_t\{C_t\}\text{(expressed in time units)}} \right\rceil + 1 \quad \textbf{and}$$

$HWinSize \in \{1, 2, \ldots, WinSize\}$ /*as chosen to be given as input to the heuristic*/

3: **for** ($\tau = 0$)

{

4: $MoveWindow \longleftarrow$ of size $WinSize$ starting from period 1 to period $\tau + WinSize$

5: $FinWindow \longleftarrow$ all periods from $WinSize + 1$ up to period T

6: **Solve** $MoveWindow \longleftarrow B_{MIP}$ and $FinWindow \longleftarrow LP_R$

}/* End of loop */

7: $\tau = 1$

8: **while** ($\tau \leq T - WinSize$) **do**

{

9: Introduce $SolnEqu_{1,i,t,\tau}$ and $SolnEqu_{2,i,t,\tau}$ corresponding to those binary variables that have had the value of 1 (up to τ) from the previous sub-problem as follows. /* It implies that all binary variables (equal to 1) are set with the corresponding $SolnEqu$'s up to period τ, and the binary variables from period 1 up to period ($\tau + WinSize$) $\in \{0, 1\}$, while solving the current sub-problem in Step 12*/:

9.1: $SolnEqu_{1,i,t,\tau} \longleftarrow$

$$\sum_{\substack{\tau'=max\{t-HWinSize;1\}\\ t\geq 2}}^{t-1} (\delta^1_{i,\tau'} + \delta^2_{i,\tau'} + \delta^3_{i,\tau'}) + \sum_{\substack{\tau'=t\\ t\geq 1}}^{min\{t+HWinSize;T\}} (\delta^1_{i,\tau'})$$

$$+ \sum_{\substack{\tau'=t\\ t\geq 1}}^{min\{t+HWinSize-1;T-1\}} (\delta^2_{i,\tau'} + \delta^3_{i,\tau'}) \geq 1$$

$\forall i \mid \delta_{i,t} = 1$ and $\forall t \leq \tau$;

Continued... Pseudocode for the Proposed Heuristic Based on MM2:CLSP-PCSC

9.2: $SolnEqu_{2,i,t,\tau} \longleftarrow$

$$\sum_{\substack{\tau'=max\{t-HWinSize;1\}\\ t\geq 2}}^{t-1} (\delta^2_{i,\tau'} + \delta^3_{i,\tau'}) + \sum_{\substack{\tau'=max\{t-HWinSize+1;1\}\\ t\geq 1}}^{min\{t+HWinSize;T\}} (\delta^1_{i,\tau'})$$

$$+ \sum_{\substack{\tau'=t\\ t\geq 1}}^{min\{t+HWinSize;T-1\}} (\delta^2_{i,\tau'} + \delta^3_{i,\tau'}) \geq 1$$

$\forall i \mid (\delta^2_{i,t} = 1)$ or $(\delta^3_{i,t} = 1)$ and $\forall t \leq \tau$.

10: $MoveWindow \longleftarrow$ of size $WinSize$ starting from period $\tau + 1$ to period $\tau + WinSize$

11: $FinWindow \longleftarrow$ all periods from $\tau + WinSize + 1$ up to period T

12: **Solve** $MoveWindow \longleftarrow B_{MIP}$, $FinWindow \longleftarrow LP_R$ and with $SolnEqu_{1,i,t,\tau}$ and $SolnEqu_{2,i,t,\tau}$ as appropriate.

13: $\tau = \tau + 1$

Note: $MoveWindow$ does not include the consideration of $SolnEqu_{1,i,t,\tau}$ and $SolnEqu_{2,i,t,\tau}$.

}/* End of while loop */

14: **end function**

Note: $\delta^1_{i,0} = \delta^2_{i,0} = \delta^3_{i,0} = 0 \quad \forall i$ (boundary conditions).

4.7 Computational Experience

In this section, the solution times of the proposed MM2:CLSP-PCSC is presented. In Table 4.3, the solution time (to obtain an optimal solution) obtained by executing the generalized MM2:CLSP-PCSC for each of the 162 sample problem instances of various product and time-period combinations is presented. All the problem instances have been executed on a Pentium 3.10 GHz Windows 7 workstation with 4.00 GB RAM, using CPLEX v12.4. It is believed that if the same problem instances are run on the latest version of CPLEX (v12.6.3), the solution times will improve. While using CPLEX, the values of the mixed integer optimality gap tolerance and integrality tolerance were set to zero, in order to obtain accuracy in results and maintain the integrality with respect to binary variables. Alternate optimal solutions are obtained for the problem instances when the default tolerance values of the CPLEX are retained.

The two tolerances which had been set to zero instead of using the default values were the integrality tolerance and the MIP tolerance. The parameter integrality tolerance sets the amount by which a computed solution value for an integer variable can violate integrality; it does not specify an amount by which CPLEX relaxes integrality. In order to avoid violation of integrality and to obtain the exact values without floating the integer solution values, it has been set it exactly to zero such that the resulting solution values are exact and rounding off the solution values need not be done manually. Sometimes having tolerance set to zero helps to compare our results with other exact methods (other mathematical models addressing the same problem), so that no variables which need not necessarily float are present in the final solution. Therefore, in order to obtain the exact values of the variables, the tolerance has been set to value 0. The parameter MIP tolerance is the gap between the best integer objective and the objective of the best node remaining. When this difference falls below the value of this parameter, the mixed integer optimization is stopped. In order to stop the optimization only at the best integer objective, the parameter has been set to zero.

The following settings are assumed while performing the computations: The time-dependent setup costs (mu/setup) of the products range from 5 mu/setup to 60 mu/setup; the time-dependent holding costs (mu/period/unit product carried over) of the products range from 1 mu/period/unit product carried over to 5 mu/period/unit product carried over; the setup time (time units) of the products (per setup) ranges from 5 time units to 460 time units and the backorder cost (mu/period/unit product backordered) of all products is assumed to be 5,000,000 mu/period/unit product backordered (which implies no backorder). Throughout this work 100 time units correspond to one time period (t) and the number of time units required for producing one unit of product i (a_i) is equal to 1 time unit per unit of product i.

We present in Table 4.4, the solution time (to obtain the optimal solution) obtained by executing the mathematical models MM1:CLSP-PCSC and MM2:CLSP-PCSC for sample problem instances of various product and time-period combinations. The MM1:CLSP-PCSC is better in terms of computational experience for most of the instances compared to MM2:CLSP-PCSC. However, when we compare the heuristics developed with respect to both the mathematical models, the heuristic developed with respect to MM1:CLSP-PCSC can be easily applied when identical capacity is present across periods. However, in reality the capacity across periods may be varying. When non-identical capacity is present across periods, for allowing the shift of setup/production for more periods ahead of or after the current time period, the extension of the heuristic based on MM1:CLSP-PCSC becomes tedious. The heuristic with respect to MM1:CLSP-PCSC allows only one left shift or right shift of setup/production ahead of or after the current time period. However, the heuristic developed with respect to MM2:CLSP-PCSC allows shift up to the end of the time horizon and can be easily adapted to unequal capacity across periods. Additionally, the heuristic developed with respect to MM2:CLSP-PCSC gives similar computational experience to the MM1:CLSP-PCSC.

Therefore, the mathematical model and heuristic developed with respect to MM2:CLSP-PCSC are more generalized that it can be applied to various problem sizes and can easily be extended for unequal capacities across periods. The mathematical model MM2:CLSP-PCSC gives comparable solution times, similar to MM1:CLSP-PCSC and the heuristic developed with respect to MM2:CLSP-PCSC is easily adaptable for unequal capacities across periods. Here the number of left shifts and right shifts that can be allowed in the heuristic is equal to the value of the time horizon. *Note:* The number of active and real variables in MM1:CLSP-PCSC and MM2:CLSP-PCSC is listed in Table 4.5.

Table 4.3: Computational time (in sec.) of the various problem instances for the MM2:CLSP-PCSC

Description	Number of Products \ Number of Periods	4	6	8	10	12	14	16	18	20
Instance 1	4	0.03	0.06	0.11	0.20	2.50	0.50	4.15	14.82	11.67
Instance 2		0.05	0.06	0.67	0.61	0.90	0.61	0.77	5.37	9.06
Average		0.04	0.06	0.39	0.41	1.70	0.56	2.46	10.10	10.37
Instance 1	6	0.06	0.09	0.52	0.48	1.00	0.92	1.28	14.49	6.54
Instance 2		0.03	0.08	0.23	0.31	0.58	1.11	4.57	2.18	35.40
Average		0.05	0.09	0.38	0.40	0.79	1.02	2.93	8.34	20.97
Instance 1	8	0.09	0.17	0.27	0.42	10.90	1.39	4.76	69.08	49.64
Instance 2		0.06	0.14	0.47	0.51	0.58	1.72	22.90	18.22	2.89
Average		0.08	0.16	0.37	0.47	5.74	1.56	13.83	43.65	26.27
Instance 1	10	0.08	0.94	0.50	0.61	1.54	5.88	3.65	6.33	11.62
Instance 2		0.06	0.19	1.48	3.64	1.40	2.81	3.26	16.13	9.73
Average		0.07	0.57	0.99	2.13	1.47	4.35	3.46	11.23	10.68
Instance 1	12	0.14	0.50	1.20	1.08	11.89	21.72	5.40	178.43	20.19
Instance 2		0.42	0.19	1.01	8.60	2.12	4.57	71.88	7.25	128.58
Average		0.28	0.35	1.11	4.84	7.01	13.15	38.64	92.84	74.39
Instance 1	14	0.33	0.73	3.67	4.51	3.57	4.62	7.69	17.77	25.43
Instance 2		0.19	0.39	2.65	2.76	2.81	12.86	13.46	32.92	1320.45
Average		0.26	0.56	3.16	3.64	3.19	8.74	10.58	25.35	672.94
Instance 1	16	0.22	3.39	5.73	1.84	3.92	9.14	16.07	174.99	71.93
Instance 2		0.20	0.31	1.11	28.52	0.89	6.27	97.41	246.68	1235.65
Average		0.21	1.85	3.42	15.18	2.41	7.71	56.74	210.84	653.79
Instance 1	18	0.42	0.90	1.79	1.82	5.21	34.41	31.50	16.05	149.73
Instance 2		0.17	1.14	2.00	2.17	7.02	55.65	12.26	19.25	516.32
Average		0.30	1.02	1.90	2.00	6.12	45.03	21.88	17.65	333.03
Instance 1	20	1.00	0.98	20.56	6.33	4.90	13.87	151.71	240.07	34.17
Instance 2		0.27	0.66	26.10	69.58	25.26	165.31	16.13	23.12	179.26
Average		0.64	0.82	23.33	37.96	15.08	89.59	83.92	131.60	106.715

Table 4.4: Computational time (in sec.) of the various problem instances for the MM1:CLSP-PCSC and MM2:CLSP-PCSC

Mathematical Models	Description		4	6	8	10	12	14	16	18	20
	Instance 1		0.03	0.08	0.09	0.19	2.71	0.86	4.96	5.65	11.84
*MM*1	Instance 2		0.03	0.06	0.45	0.64	0.76	0.61	1.78	4.09	3.82
	Average		0.03	0.07	0.27	0.42	1.74	0.74	3.37	4.87	7.85
	Instance 1	4	0.03	0.06	0.11	0.20	2.50	0.50	4.15	14.82	11.67
*MM*2	Instance 2		0.05	0.06	0.67	0.61	0.90	0.61	0.77	5.37	9.06
	Average		0.04	0.06	0.39	0.41	1.70	0.56	2.46	10.10	10.37
	Instance 1		0.05	0.09	0.59	0.53	0.72	1.08	1.79	10.17	9.08
*MM*1	Instance 2		0.03	0.06	0.22	0.30	0.61	1.14	2.59	2.89	39.78
	Average		0.04	0.08	0.41	0.42	0.67	1.11	2.19	6.53	24.43
	Instance 1	6	0.06	0.09	0.52	0.48	1.00	0.92	1.28	14.49	6.54
*MM*2	Instance 2		0.03	0.08	0.23	0.31	0.58	1.11	4.57	2.18	35.40
	Average		0.05	0.09	0.38	0.40	0.79	1.02	2.93	8.34	20.97
	Instance 1		0.06	0.13	0.22	0.42	2.73	1.68	2.51	30.87	8.92
*MM*1	Instance 2		0.05	0.13	0.38	0.45	2.53	1.70	2.64	4.76	6.72
	Average		0.06	0.13	0.30	0.44	2.63	1.69	2.58	17.82	7.82
	Instance 1	8	0.09	0.17	0.27	0.42	10.90	1.39	4.76	69.08	49.64
*MM*2	Instance 2		0.06	0.14	0.47	0.51	0.58	1.72	22.90	18.22	2.89
	Average		0.08	0.16	0.37	0.47	5.74	1.56	13.83	43.65	26.27
	Instance 1		0.05	0.45	0.31	0.72	1.30	2.42	7.53	6.80	12.84
*MM*1	Instance 2		0.05	0.17	0.50	3.34	2.14	2.59	3.93	6.04	9.88
	Average		0.05	0.31	0.41	2.03	1.59	2.51	5.73	6.42	11.36
	Instance 1	10	0.08	0.94	0.50	0.61	1.54	5.88	3.65	6.33	11.62
*MM*2	Instance 2		0.06	0.19	1.48	3.64	1.40	2.81	3.26	16.13	9.73
	Average		0.07	0.57	0.99	2.13	1.47	4.35	3.46	11.23	10..68
	Instance 1		0.09	0.25	0.53	0.78	2.34	3.32	15.38	50.17	18.17
*MM*1	Instance 2		0.20	0.17	0.51	1.75	1.87	3.53	28.27	11.00	20.64
	Average		0.15	0.21	0.52	1.27	2.11	3.43	21.83	30.59	19.41
	Instance 1	12	0.14	0.50	1.20	1.08	11.89	21.72	5.40	178.43	20.19
*MM*2	Instance 2		0.42	0.19	1.01	8.60	2.12	4.57	71.88	7.25	128.58
	Average		0.28	0.35	1.11	4.84	7.01	13.15	38.64	92.84	74.39
	Instance 1		0.17	0.39	0.97	1.31	14.09	5.82	7.50	14.31	18.53
*MM*1	Instance 2		0.26	0.23	0.72	4.60	3.48	4.46	13.38	16.68	445.77
	Average		0.22	0.31	0.85	2.96	8.79	5.14	10.44	15.50	232.15
	Instance 1	14	0.33	0.73	3.67	4.51	3.57	4.62	7.69	17.77	25.43
*MM*2	Instance 2		0.19	0.39	2.65	2.76	2.81	12.86	13.46	32.92	1320.45
	Average		0.26	0.56	3.16	3.64	3.19	8.74	10.58	25.35	672.94
	Instance 1		0.11	0.69	1.45	1.39	3.38	5.60	10.02	20.36	497.80
*MM*1	Instance 2		0.19	0.28	1.01	4.26	1.14	5.60	10.92	36.24	29.60
	Average		0.15	0.49	1.23	2.83	2.26	5.60	10.47	28.30	263.70
	Instance 1	16	0.22	3.39	5.73	1.84	3.92	9.14	16.07	174.99	71.93
*MM*2	Instance 2		0.20	0.31	1.11	28.52	0.89	6.27	97.41	246.68	1235.65
	Average		0.21	1.85	3.42	15.18	2.41	7.71	56.74	210.84	653.79
	Instance 1		0.69	0.33	1.20	3.03	5.46	19.84	13.51	19.72	56.86
*MM*1	Instance 2		0.13	0.67	1.00	4.38	9.45	7.21	16.26	18.75	94.02
	Average		0.41	0.5	1.1	3.71	7.46	13.53	14.89	19.24	75.44
	Instance 1	18	0.42	0.90	1.79	1.82	5.21	34.41	31.50	16.05	149.73
*MM*2	Instance 2		0.17	1.14	2.00	2.17	7.02	55.65	12.26	19.25	516.32
	Average		0.30	1.02	1.90	2.00	6.12	45.03	21.88	17.65	333.03
	Instance 1		0.61	0.47	1.36	2.82	4.74	13.17	38.61	31.82	102.20
*MM*1	Instance 2		0.27	0.44	2.21	10.33	5.32	8.64	13.00	19.45	96.03
	Average		0.44	0.46	1.79	6.58	5.03	10.62	25.81	25.64	99.12
	Instance 1	20	1.00	0.98	20.56	6.33	4.90	13.87	151.71	240.07	34.17
*MM*2	Instance 2		0.27	0.66	26.10	69.58	25.26	165.31	16.13	23.12	179.26
	Average		0.64	0.82	23.33	37.96	15.08	89.59	83.92	131.60	106.715

Table 4.5: Number of active binary variables in a given time horizon

MM1:CLSP-PCSC		MM2:CLSP-PCSC	
Variable Notation	Maximum number of active binary variables	Variable Notation	Maximum number of active binary variables
Integer Variables			
$\delta^1_{i,t}$	$\leq N \times T$	$\delta^1_{i,t}$	$\leq N \times T$
$\delta^2_{i,t}, \delta^3_{i,t}$	$\leq T$	$\delta^2_{i,t}, \delta^3_{i,t,t'}, \delta^4_{i,t,t'}$	$\leq T$
$\Delta^1_{i,t,t'}$	$\leq N \times T$	$\Delta^1_{i,t,t'}$	$\leq N \times T$
$\Delta^2_{i,t,t'}, \Delta^3_{i,t,t'}, \Delta^4_{i,t,t'}$	$\leq T$	$\Delta^2_{i,t,t'}, \Delta^3_{i,t,t',t''}, \Delta^4_{i,t,t',t''}$	$\leq T$
$\Omega^3_{i,t,t'}$	$\leq 2 \times T$		
$\Omega^4_{i,t,t'}$	$\leq T$		
Real Variables			
$I_{i,t}$	$= N \times T$	$I_{i,t}$	$= N \times T$
$B_{i,t}$	$= N \times T$	$B_{i,t}$	$= N \times T$
$s^1_{i,t}$	$\leq N \times T$	$s^1_{i,t}$	$\leq N \times T$
$s^2_{i,t}, s^3_{i,t,t'}, s^4_{i,t,t'}$	$\leq T$	$s^2_{i,t}, s^3_{i,t,t',t''}, s^4_{i,t,t',t''}$	$\leq T$
$X^1_{i,t,t'}$	$\leq N \times T$	$X^1_{i,t,t'}$	$\leq N \times T$
$X^2_{i,t,t'}, X^3_{i,t,t'}, X^4_{i,t,t'}$	$\leq T$	$X^2_{i,t,t'}, X^3_{i,t,t',t''}, X^4_{i,t,t',t''}$	$\leq T$

4.8 Summary

The mathematical model (MM2:CLSP-PCSC) proposed in this chapter helps to address situations in process industries such as the presence of long setup times, and the setup of a product and its consecutive production are carried across periods as well as real-life situations such as production immediately after setup and uninterrupted production carryover across multiple time periods. In addition, the setup costs may be time independent or dependent. A comprehensive heuristic is also proposed based on MM2:CLSP-PCSC. In Chap. 3, the MM1:CLSP-PCSC is better in terms of computational experience for most of the instances compared to MM2:CLSP-PCSC proposed in this chapter. However, when we compare the heuristics developed with respect to both the mathematical models, the heuristic developed with respect to MM1:CLSP-PCSC can be easily applied when identical capacity is present across periods. However, in reality the capacity across periods may be varying. When non-identical capacity is present across periods, for allowing the shift of setup/production for more periods ahead of or after the current time period, the extension of the heuristic based on MM1:CLSP-PCSC becomes tedious. The heuristic with respect to MM1:CLSP-PCSC allows only one left shift or right shift of setup/production ahead of or after the current time period. However, the heuristic developed with respect to MM2:CLSP-PCSC allows shift up to the end of the time horizon and can be easily adapted to unequal capacity across periods.

Therefore, the mathematical model and heuristic developed with respect to MM2:CLSP-PCSC are more generalized and give similar computational experience to the MM1:CLSP-PCSC. It is more generalized because it can be easily applied to various problem sizes and can easily be extended for unequal capacities across periods. The mathematical model MM2:CLSP-PCSC gives comparable solution times, similar to MM1:CLSP-PCSC and the heuristic developed with respect to MM2:CLSP-PCSC is easily adaptable for unequal capacities across periods. Here the number of left shifts and right shifts that can be allowed in the heuristic is equal to the value of the time horizon.

CHAPTER 5

Capacitated Lot Sizing Problem with Production Carryover and Setup Crossover Across Periods Assuming Sequence-Dependent Setup Times and Setup Costs (CLSP-SD-PCSC): Mathematical Models for Process Industries

5.1 Introduction and Problem Definition

In Chaps. 3 and 4, mathematical models have been proposed for the capacitated lot sizing problem with production carryover and setup crossover across periods. Heuristics based on both the mathematical models have also been proposed. The models and heuristics address real-life situations in process industries such as production immediately after setup and uninterrupted production carryover across periods.

In process industries, products may have short or long setup times, and the setup cost and setup time may be sequence independent or sequence dependent. This chapter examines the capacitated lot sizing problem (CLSP) addressing situations present in process industries, with sequence-dependent setup costs and setup times. In any industry, in order to produce the product the machine has to be set up to make it ready for production. The setup time of the machine not only depends on the time to set up the particular product, it can also depend upon the previous product for which the machine was set up for, and other procedures which need to be undertaken before setting up the next product. Let us consider i' and i as two products which need to be set up on a single machine successively. When the setup time of a product on the machine depends on the time taken to setup product i after setting up the previous product i', the setup is called as sequence-dependent setup and the corresponding time taken for setting up the product is called as sequence-dependent setup time. The corresponding cost involved in setting up product i after setting up product i' is called sequence-dependent setup cost. Sequence-dependent setup times are considered when the setup time of a product has to be anticipated not only based on the time to setup product i, but also due to the product setup previously (i'). This is so because the setup of product i cannot be performed as soon as the machine is free after producing i', because of certain factors such as cleaning of the machine and variable fixing time of the machine tools to process the next product.

R. Ramya et al., *Capacitated Lot Sizing Problems in Process Industries*, https://doi.org/10.1007/978-3-030-01222-9_5

Sequence-dependent setup times are present in industries such as automobile, semiconductor, printed circuit board, flexible manufacturing and plastic manufacturing industries. The production planning problem in a numerically controlled milling machine in which specific tools and holders need to be mounted considers lot sizing along with sequence-dependent setup costs and times (Haase and Kimms 2000). Some of the re-entrant flow shops (here products are processed twice on the same machine), i.e., manufacturers of printed circuit boards, mirrors and semiconductors also have sequence-dependent setup times (Jeong and Kim 2014). Separable sequence-dependent setup times are present in flexible manufacturing systems (FMS) where the machine can be set up well in advance before the product which is to be processed is available, after the processing of the previous product has been completed (Abdelmaguid 2015).

Some of the process industries which consider sequence-dependent setup times are dyeing, packaging, glass container manufacturing, abrasive manufacturing, paper, printing, textile and chemical processing industry. Sequence dependency is considered in a steel mini-mill while doing the slab casting process. Here if a narrow width of slab is produced after a wide one, a small amount of setup time is incurred, and if the opposite sequence is produced, a large amount of setup time is incurred (Park et al. 2002). Sequence-dependent setup times exist for a company which produces sandpaper rolls of different grades or roughness. The production process starts with the production of large rolls of paper that are later cut into different sizes based on the customer orders. Here the setup initiated is sequence dependent because different types of glue are used in the production process. The cleaning up of the machine after each batch also results in significant setup times. Processes with sequence-dependent setups include the stamping operation in plastic manufacturing and roll slitting in the paper industry (Eren and Güner 2006). In the dyeing industry, the setup times vary when different colours are changed. It takes less time to change shades of some colours, for example, from a pale colour to a dark colour (small setup time), but it takes more time to clean the dyeing vessel. At times the setup time becomes greater than or equal to the operation time in a dyeing industry (long setup time) due to which sequence-dependent setup costs and setup times are incorporated while doing the production planning. Sequence-dependent setup times and setup costs are also predominantly found in industries having multi-functional abilities. In an industry manufacturing chemical compounds, the cleaning of the machine depends upon the chemical previously processed and the chemical to be processed next. In a printing industry, the cleaning of the machine depends on the colour of the ink and the size of the paper for which the previous setup was done (Kwak and Jeong 2011). Lang and Shen (2011) addressed the production planning problem of a manufacturer of plastic sheets (of several variants) meant for car wind shields. These plastic sheets form an internal layer between the two layers of the wind shield glass. Sequence-dependent setup times and setup costs occur while manufacturing the plastic sheets, while changing from one foil variant to another because of the various attributes the foils differ in, such as width, thickness, colour, length, material, surface of the foil (course or even) and prevention of adhering (foil is cooled or an "anti-adhesive paper" is added on the roll to prevent the foil from adhering). In a beverage industry, switching between products due to difference in container size or shape or type of liquid leads to sequence-dependent setup times and setup costs. In a textile industry, sequence-dependent setup times and setup costs arise while changing the fibre blend or while doing adjustments in the yarn machines (Guimarães et al. 2014).

There are also semi-continuous process industries which involve sequence-dependent setup costs and setup times. Almada-Lobo et al. (2007) and Almada-Lobo and James (2010) considered the production process in a glass container manufacturing industry which involved the production of glass (continuous part) and container (discrete part). The melting of the raw materials such as silica sand, sodium oxide from soda ash, calcium oxide from limestone/dolomite and feldspar corresponds to the continuous manufacturing process wherein these materials are passed through a furnace which is heated up to 1500 °C. Sequence-dependent setup time and a proportional setup cost is considered during a product changeover in the moulding machine which processes the glass. This is because the machine is fed with molten glass from the furnace continuously, wherein the gobs are discarded and melted down again in the furnace due to which a huge amount of energy is wasted, thereby increasing the production cost. Therefore, sequence-dependent setup time and a proportional setup cost is considered for faster and fewer product changes in order to lower the energy costs. CLSP with sequence-dependent setup costs and setup times was considered by several researchers such as Kang et al. (1999), Clark and Clark (2000), Timpe (2002) and Lang and Shen (2011). CLSP with production carryover across periods and sequence-dependent setup costs and times was considered by Haase (1996), Haase and Kimms (2000), Almada-Lobo et al. (2007), Kovács et al. (2009), Almada-Lobo and James (2010), Mohammadi et al. (2010), Kwak and Jeong (2011), Shim et al. (2011), Clark et al. (2014) and Guimarães et al. (2014). In this chapter, mathematical models for the capacitated lot sizing problem with sequence-dependent setup costs and setup times are proposed, considering production carryover and setup crossover across periods, with backorders allowed. The mathematical models which have been developed for the CLSP under study consider the presence of long setup times (in the order of hours/shifts/days or weeks) which are present in process industries. The mathematical models can also address industrial situations where the setup time of a product is less than the capacity of a period, thereby allowing setup splitting between two consecutive periods in order to facilitate the effective utilization of capacity. The setup splitting/crossover mainly depends upon the duration of setup time of a product. Referring to the benchmark papers that are related to our study (e.g. Sung and Maravelias (2008) and Belo-Filho et al. (2013)), we find that the setup time may extend beyond one time bucket and the setup time is comparable or perhaps more than the production run time (i.e. the time for setup may extend beyond a shift or a day. This is especially true in case of process industries such as chemical, sugar, cement and metallurgical industries. The situation where the setup time is possibly more than the production run time holds true in case of high value and process intensive industries such as pharmaceutical industries where the production quantity per unit time fetches high revenue in comparison to the cost incurred with respect to long setup duration (due to high technology intensive setups associated with process controls and monitoring while setting up the process for manufacturing, for the production of end products associated with high cost raw materials and high revenue/ unit sales end products). However, the proposed model in this chapter is not constrained by the consideration of long setup products. The model is flexible enough to handle the process industries with small bucket setups and long bucket production runs or the scenario with large bucket setups and small production runs or a mixture of both. In other words, the proposed mathematical models are flexible enough to handle or address situations in the conventional process industries such as cement and sugar industries (associated with small bucket setups and long bucket production runs), large bucket setups and small bucket production runs (associated with highly technological intensive big bucket setups and small bucket production runs such as those in highly specialized pharmaceutical processes) or a mixture of scenarios in a single process industry. Also, depending upon the industry the definition of a period may vary. The mathematical models can also address industrial situations where setup times of products are sequence dependent. These models mainly address real-life situations present in process industries where the presence of idle time is not allowed between the setup of a product and its consecutive production, and when the production carryover of the product takes place across periods. These real-life considerations are brought into the proposed model because in certain industries, the production should start immediately after setup and no break in production during the course of production of a product is allowed. Such situations are present in industries such as hot rolling process and heat treatment (in manufacturing industries), glass manufacturing, chemical manufacture, cement manufacture, sugar industry and pharmaceuticals. Considering the above real-life situations in process industries along with the presence of long/short setup times, and sequence-dependent setup costs and times, can be a significant contribution to the CLSP literature.

5.2 Basic Assumptions of the Proposed Mathematical Models (MM1: CLSP-SD-PCSC and MM2:CLSP-SD-PCSC)

- A single machine is considered in the problem.
- Multiple products can be produced on the single machine and each product is made up of a single level.
- Time unit is discrete and the time horizon considered is finite.
- Each product is associated with a setup cost when set up on the machine, and it consumes time for setup.
- Backorders are allowed but lost sales are not permitted.
- The capacity of the machine during a given period is assumed in time units and it may vary from period to period.
- The capacity of the machine per period is consumed by the setup time and the production time of the products. Idle time on the machine can also be present.
- If excess capacity is left over on the machine in a period after production in period t, it may be used to setup the product to be produced in the next period. If the setup is not over, this setup may be continued to some future period t', where $t' > t$. In this work this aspect is called setup crossover.
- The excess quantity produced of a product can be stored and this incurs a holding cost, except in the last period where all the units in the inventory have to be consumed.
- Production of a product may extend over any number of periods subject to demand and capacity constraints. In this work this aspect is called production carryover.

- Almost one setup of a given product i can be initiated in the given time period t. It means that the carryover of a setup is permitted from one of the previous time periods to end in the present period, and the initiation of setup of that product in that given time period is permitted subsequently after the production of any other product in-between (i.e. setup of the same product cannot be initiated twice within a given time period).

- Production of the same product cannot occur more than once within a given time period i.e., the production of the product cannot be implemented due to either by idling of the production resources or due to the production of any other product in-between.

- The triangular inequality with respect to setup times of products i, i' and i'' is also assumed (i.e. $ST_{i,i'} + ST_{i',i''} \geq ST_{i,i''}$ and $ST_{\phi,i} + ST_{\phi,i'} \geq ST_{\phi,i'}$), and so is the case with respect to setup costs of products i, i' and i''.

5.3 Mathematical Model 1 (MM1:CLSP-SD-PCSC)

The parameters, decision variables and the generalized mathematical model MM1:CLSP-SD-PCSC are presented in this section.

5.3.1 Parameters/Indices

N	number of products
T	number of time periods
t	a given time period
i	product
$SC_{\phi,i}$	sequence-dependent setup cost incurred in the machine for the first product i setup in period 1
$SC_{i',i}$	sequence-dependent setup cost incurred, when the machine is set up from product i' to product i
b_i	backorder cost per period per unit of product i
h_i	holding cost per period per unit of product i
$ST_{\phi,i}$	sequence-dependent setup time for the first product i setup in the machine in period 1
$ST_{i',i}$	sequence-dependent setup time when the machine is set up from product i' to product i
a_i	number of time units required for producing one unit of product i
C_t	capacity of the machine in period t (in time units)
$d_{i,t}$	demand for product i in period t
M	a large value (here the value of *M* assigned for every problem instance is the largest setup time among all products (i.e. $M = max(ST_{\phi,i}\,\forall i;\, ST_{i',i}\,\forall i'$ and $\forall i$) is considered)
$\mathcal{E}$	smallest unit of time
$\mathcal{E}_d$	unit of smallest quantity of production

5.3.2 Decision Variables

Variable	Description
$\delta^1_{i,t}$	An indicator (binary) variable that takes value 1 if a complete setup is done for product i in period t with the production starting in period t; 0 otherwise.
$\Delta^1_{i,t,t'}$	An indicator (binary) variable that takes value 1: it corresponds to a possible production carryover from period t to period t' ($t \leq t' \leq T$), due to the setup of product i started and finished in period t, with no intermittent setup of any other product; 0 otherwise.
$\delta^2_{i,t}$	An indicator (binary) variable that takes value 1 if the setup of product i is started and completed exactly at the end of period t, followed by its production starting in period $t+1$; 0 otherwise.
$\Delta^2_{i,t,t'}$	An indicator (binary) variable that takes value 1: it corresponds to a possible production carryover from period t' to period $t'+1$ ($t+1 \leq t' \leq T$), due to the end-of-period setup of product i in period t, with no intermittent setup of any other product; 0 otherwise.
$\delta^3_{i,t}$	An indicator (binary) variable that takes value 1 if the setup of product i is commenced in period t and is carried over across periods, and is completed in some period t' ($t' = t+1, t+2, .., T$); 0 otherwise.
$\Omega^3_{i,t,t'}$	An indicator (binary) variable that takes value 1 if the setup of product i is initiated in period t and is present in period t' ($t \leq t' \leq T$), with the setup of product i ending in a period later than period t, but not exactly at the end of that period; 0 otherwise.
$\Omega^4_{i,t,t'}$	An indicator (binary) variable that takes value 1 if the setup of product i is initiated in period t and is present in period t' ($t \leq t' \leq T-1$), with the setup of product i ending in a period later than period t and setup getting completed exactly at the end of that period; 0 otherwise.
$\Delta^3_{i,t,t'}$	An indicator (binary) variable that takes value 1: it corresponds to a possible production carryover from period t' to period $t'+1$ ($t+1 \leq t' \leq T$), with the setup of product i (having started in period t) ending in a period later than period t, but not exactly at the end of that period, and with no intermittent setup of any product during the production of product i; 0 otherwise.

Variable	Description
$\Delta^4_{i,t,t'}$	An indicator (binary) variable that takes value 1: it corresponds to a possible production carryover from period t' to period $t'+1$ ($t+2 \leq t' \leq T$), with the setup of product i (having started in period t) ending in a period later than period t and setup getting completed exactly at the end of that period, and with no intermittent setup of any product during the production of product i; 0 otherwise.
$I_{i,t}$	Inventory of product i at the end of period t.
$B_{i,t}$	Backorder quantity of product i at the end of period t.
$s^1_{i,t}$	Setup time of product i in period t that takes the value of $ST'_{i,t}$, and associated with $\delta^1_{i,t}$.
$s^2_{i,t}$	Setup time of product i in period t that takes the value of $ST'_{i,t}$, and associated with $\delta^2_{i,t}$.
$s^3_{i,t,t'}$	Setup time of product i in period t' that takes the value of $ST'_{i,t}$ due to its setup started in period t, and associated with $\Omega^3_{i,t,t'}$.
$s^4_{i,t,t'}$	Setup time of product i in period t' that takes the value of $ST'_{i,t}$ due to its setup started in period t, and associated with $\Omega^4_{i,t,t'}$.
$X^1_{i,t,t'}$	Production quantity of product i in period t' (due to its setup starting and ending within period t), with $1 \leq t \leq T$ and $t \leq t' \leq T$, and associated with $\Delta^1_{i,t,t'}$.
$X^2_{i,t,t'}$	Production quantity of product i in period t' (due to its setup started in period t and completed exactly at the end of that period), with $1 \leq t \leq T-1$ and $t+1 \leq t' \leq T$, and associated with $\Delta^2_{i,t,t'}$.
$X^3_{i,t,t'}$	Production quantity of product i in period t' (due to its setup starting in period t and ending in a later period but not at the end of that period), with $1 \leq t \leq T-1$ and $t+1 \leq t' \leq T$, and associated with $\Delta^3_{i,t,t'}$.
$X^4_{i,t,t'}$	Production quantity of product i in period t' (due to its setup starting in period t and ending at the end of a later period), with $1 \leq t \leq T-2$ and $t+2 \leq t' \leq T$, and associated with $\Delta^4_{i,t,t'}$.
$del_{\phi,i,1}$	An indicator (binary) variable which tracks product i that has been setup with respect to the beginning of time period 1.
$del_{0,i,t}$	An indicator (binary) variable which tracks product i that has been setup last, with respect to the beginning of time period t.
$del_{i,0,t}$	An indicator (binary) variable which tracks product i that has been setup last, with respect to the end of time period t.
$\xi'_{\phi,i,1}$	An indicator (binary) variable which indicates the initiation of setup of product i as the first setup in period 1.

Variable	Description
$\xi_{i',i,t}$	An indicator (binary) variable which indicates the setup of product i in period t, with the setup of both products i' and i being initiated in this period. Therefore it is evident that at least one of $\xi_{i',i,t}$ cannot exist in period t without being preceded by $\xi'_{i'',i',t}$. In other words, the first possible setup initiated in this period is denoted by $\xi'_{i'',i',t}$ (in general followed by subsequent possible initiated setups denoted by $\xi_{i',i,t}$).
$\xi'_{i,i',t}$	An indicator (binary) variable which indicates the initiation of setup of product i' as the first setup in period t, with the setup of product i' preceded by the setup of product i.
$delta_t$	An indicator (binary) variable which takes the value 1, if the setup of at least one of any products is initiated in period t (indicated by $delta_t = 1$). It also means that no carryover of either setup or production of any product, with its setup having been initiated in any period up to period $t-1$, can be carried over to period $t+1$ and later. This is so because if any product's setup is initiated in period t, then either the production or setup carryover with the setup initiated up to period $t-1$ which can carried over to period $t+1$ and later is not feasible.
$U_{i,t}$	An auxiliary variable that assigns product i in period t; associated with $\xi_{i',i,t}$ to avoid sub-tours.
$ST'_{i,t}$	A variable that is assigned the value of setup time of product i which is initiated for its setup in period t.

5.3.3 Mathematical Model 1 (MM1:CLSP-SD-PCSC)

In this section, the mathematical model (MM1:CLSP-SD-PCSC) is presented, with an objective of minimizing the sequence-dependent setup costs, holding costs and backorder costs of all products across all time periods. The proposed mathematical model assumes sequence-dependent setup costs and setup times. The triangular inequality with respect to setup times of products i, i' and i'' is also assumed (i.e. $ST_{i,i'} + ST_{i',i''} \geq ST_{i,i''}$ and $ST_{\phi,i} + ST_{\phi,i'} \geq ST_{\phi,i''}$), and so is the case with respect to setup costs of products i, i' and i''. This mathematical model helps to address production situations in process industries. They address situations such as the presence of long setup times (sequence dependent), production starting immediately after the product's setup and uninterrupted production carryover across periods. In this mathematical model, three binary variables are used to track three types of setup, i.e., a complete setup done (started and finished) anywhere in period t (but not exactly at the end of period t), an end-of-period setup and a setup crossover which is initiated in period t due to the presence of long setup times. This setup crossover indicator is in turn linked with two other binary variables. One variable indicates a setup crossing over across a number of periods with the setup initiated anywhere in period t and ending in the middle of some future period t', and another variable indicates a setup crossing over across a number of periods with the setup initiated anywhere in period t and completed exactly at the end of period t'. These binary variables which indicate a product's setup are in turn linked with production carryover indicator variables that help to track the time period in which the product's setup is initiated and completed, and the time period in which the corresponding production is carried out. Production variables corresponding to these production carryover indicator variables are also present which help to determine the production time (here, the production quantity is measured in time units). There are variables which determine the setup time of a product in a period. When the setup of product i crosses over a number of periods, the time taken to setup product i is also split across these periods.

In order to address the presence of sequence-dependent setup times, a (binary) variable which indicates the initiation of setup of product i' as the first setup in period 1 is introduced. Similarly, an indicator (binary) variable which indicates the initiation of setup of product i' as the first setup in period t (where $2 \leq t \leq T$), with the setup of product i' preceded by the setup of product i'', is also introduced. In order to indicate the subsequent possible setups in all periods, another binary variable is introduced. This variable indicates the setup of product i in period t, with the setup of both products i' and i being initiated in the same period. Also, this variable does not exist without the first setup being initiated in that period. Binary variables are also introduced to track the product that has been setup last, with respect to the beginning of every time period and with respect to the end of every time period. The setup variables (complete setup, end-of-period or setup crossover) are linked with these setup initiating variables as well as the tracking variables. Flow balance constraints are introduced to ensure proper inflow and outflow of setups in a period (i.e. if setup of product i' takes place in period t, after setting up product i'' in period t or in any earlier period, then the next setup of product i has to take place only from product i'), and a sub-tour elimination constraint is introduced to remove any sub-tours that can be formed. Constraints are introduced to ensure that a production carryover of product i cannot take place from period t to period t' ($t' > t$) if a setup of product i' succeeds the setup of product i in period t. Similarly constraints are introduced to ensure that setup of product i (setup taking place after the first setup in every period t) cannot take place in period t without setup of product i' (first setup in every period t). These two setups are handled by separate binary indicator variables. There are also constraints which assign the value of the sequence-dependent setup time for product i in period t, when a setup of product i takes place from product i' in period t. It is also ensured that if the setup of at least 1 product is initiated in period t, then it means that no carryover of either setup or production of any product, with its setup having been initiated in any period up to $t - 1$, can be carried over to period $t + 1$ or later. This is so because if any product's setup is initiated in period t, then neither the production nor setup carryover with the setup initiated up to period $t - 1$ and carried over to period $t + 1$ and later, is feasible. There are also constraints to ensure that production starts immediately after setup and uninterrupted production takes place across periods. Through this mathematical model, it is ensured that the demand for all products is satisfied across the entire time horizon with the condition that the production time and the setup time of the products setup in a period do not exceed the capacity limitations (measured in time units) of that period.

Objective Function:

$$\text{Min } Z = \sum_{i=1}^{N}(SC_{\phi,i} \times \xi'_{\phi,i,1}) + \sum_{\substack{i'=1 \\ i'\neq i}}^{N}\sum_{i=1}^{N}\sum_{t=2}^{T}(SC_{i',i} \times \xi'_{i',i,t}) + \sum_{\substack{i'=1 \\ i'\neq i}}^{N}\sum_{i=1}^{N}\sum_{t=1}^{T}(SC_{i',i} \times \xi_{i',i,t}) + \sum_{i=1}^{N}\sum_{t=1}^{T} h_i I_{i,t} + \sum_{i=1}^{N}\sum_{t=1}^{T} b_i B_{i,t} \quad (5.1)$$

Subject to the Following:

/* Constraints (5.2)–(5.5) indicate that only one product should be tracked (concerning its setup and production) with respect to the beginning and end of every period t */

$$\sum_{i=1}^{N} del_{\phi,i,1} = 1. \quad (5.2)$$

$$\sum_{i=1}^{N} del_{i,0,1} = 1. \quad (5.3)$$

$$\sum_{i=1}^{N} del_{0,i,t} = 1, \quad t=2,3,..,T. \quad (5.4)$$

$$\sum_{i=1}^{N} del_{i,0,t} = 1, \quad t=2,3,..,T. \quad (5.5)$$

/* Constraint (5.6) indicates that the product tracked last (concerning its setup and production) with respect to period $t-1$ is equal to the product tracked last with respect to the beginning of period t */

$$del_{0,i,t} = del_{i,0,t-1} \quad \forall i \text{ and } t=2,3,..,T. \quad (5.6)$$

/* Constraint (5.7) helps to track the product setup with respect to the beginning of period 1 */

$$del_{\phi,i,1} \leq \delta^1_{i,1} + \delta^2_{i,1} + \delta^3_{i,1} \quad \forall i. \quad (5.7)$$

/* Constraints (5.8)–(5.16) help to track the product that has been setup last with respect to the end of period t */

$$del_{i,0,t} \leq 2 - (\delta^2_{i,t} + \delta^3_{i,t}) \quad \forall i \text{ and } \forall t. \quad (5.8)$$

$$del_{i,0,t} \geq \delta^2_{i,t} + \delta^3_{i,t} \quad \forall i \text{ and } \forall t. \quad (5.9)$$

$$del_{i,0,1} \leq del_{\phi,i,1} + \sum_{\substack{i'=1 \\ i'\neq i}}^{N}(\delta^1_{i',1} + \delta^2_{i',1} + \delta^3_{i',1}) \quad \forall i. \quad (5.10)$$

$$del_{i,0,1} \geq del_{\phi,i,1} - \sum_{\substack{i'=1 \\ i' \neq i}}^{N} (\delta^1_{i',1} + \delta^2_{i',1} + \delta^3_{i',1}) \qquad \forall i. \tag{5.11}$$

$$del_{i,0,t} \leq del_{0,i,t} + \sum_{i'=1}^{N} (\delta^1_{i',t} + \delta^2_{i',t} + \delta^3_{i',t}) \qquad \forall i \text{ and } t=2,3,..,T. \tag{5.12}$$

$$del_{i,0,t} \geq del_{0,i,t} - \sum_{i'=1}^{N} (\delta^1_{i',t} + \delta^2_{i',t} + \delta^3_{i',t}) \qquad \forall i \text{ and } t=2,3,..,T. \tag{5.13}$$

$$\delta^2_{i,t} + \delta^3_{i,t} \leq del_{i,0,t} \qquad \forall i \text{ and } t=1,2,..,T\text{-}1. \tag{5.14}$$

$$del_{i,0,1} \leq del_{\phi,i,1} + \delta^1_{i,1} + \delta^2_{i,1} + \delta^3_{i,1} \qquad \forall i. \tag{5.15}$$

$$del_{i,0,t} \leq del_{0,i,t} + \delta^1_{i,t} + \delta^2_{i,t} + \delta^3_{i,t} \qquad \forall i \text{ and } t=2,3,..,T. \tag{5.16}$$

/* Constraints (5.17)–(5.19) assign the product setup with respect to the beginning of every period t to the indicator variable */

$$\xi'_{\phi,i,1} = del_{\phi,i,1} \qquad \forall i. \tag{5.17}$$

$$\sum_{\substack{i'=1 \\ i' \neq i}}^{N} \xi'_{i,i',t} \leq del_{0,i,t} \qquad \forall i \text{ and } t=2,3,..,T. \tag{5.18}$$

$$\sum_{\substack{i'=1 \\ i' \neq i}}^{N} \xi'_{i',i,t} \leq 1 - del_{0,i,t} \qquad \forall i \text{ and } t=2,3,..,T. \tag{5.19}$$

/* Constraint (5.20) indicates that at least one product should be set up as the first setup in period 1 */

$$\sum_{i=1}^{N} \xi'_{\phi,i,1} = 1. \tag{5.20}$$

/* Constraint (5.21) indicates that the binary variable takes the value 1 if at least one product i' is set up as the first setup in period t */

$$\sum_{\substack{i=1 \\ i \neq i'}}^{N} \sum_{i'=1}^{N} \xi'_{i,i',t} \leq 1, \qquad t=2,3,..,T. \tag{5.21}$$

/* Constraints (5.22)–(5.24) assign the value 1 to the indicator variable when product i which is tracked for its setup with respect to the end of every period t */

$$del_{i,0,1} \leq \xi'_{\phi,i,1} + \sum_{\substack{i'=1 \\ i' \neq i}}^{N} \xi_{i',i,1} \qquad \forall i. \tag{5.22}$$

$$del_{i,0,t} \leq 1 - \sum_{\substack{i'=1 \\ i' \neq i}}^{N} \xi_{i,i',t} \qquad \forall i \text{ and } \forall t. \tag{5.23}$$

$$del_{i,0,t} \leq del_{0,i,t} + \sum_{\substack{i'=1 \\ i' \neq i}}^{N} \xi'_{i',i,t} + \sum_{\substack{i'=1 \\ i' \neq i}}^{N} \xi_{i',i,t}, \qquad \forall i \text{ and } t=2,3,..,T. \tag{5.24}$$

/* Constraints (5.25)–(5.28) link the setup of product i denoted by $\delta^1_{i,t}$, $\delta^2_{i,t}$ and $\delta^3_{i,t}$, with the indicator variable which assigns the setup of product i initiated in period t preceded by product i' */

$$\sum_{\substack{i'=1\\ i'\neq i}}^{N} \xi_{i,i',t} \leq \delta^1_{i,t} \qquad \forall i \text{ and } \forall t. \tag{5.25}$$

$$\sum_{\substack{i'=1\\ i'\neq i}}^{N} \xi_{i,i',t} \leq 1 - \left(\delta^2_{i,t} + \delta^3_{i,t}\right) \qquad \forall i \text{ and } \forall t. \tag{5.26}$$

$$\xi'_{\phi,i,1} + \sum_{\substack{i'=1\\ i'\neq i}}^{N} \xi_{i',i,1} = \delta^1_{i,1} + \delta^2_{i,1} + \delta^3_{i,1} \qquad \forall i. \tag{5.27}$$

$$\sum_{\substack{i'=1\\ i'\neq i}}^{N} \left(\xi'_{i',i,t} + \xi_{i',i,t}\right) = \delta^1_{i,t} + \delta^2_{i,t} + \delta^3_{i,t} \qquad \forall i \text{ and } t\text{=2,3,..,}T. \tag{5.28}$$

/* Constraints (5.29)–(5.32) represent the flow balance constraints with respect to setup carryover tracking and the current period's setup tracking */

$$\xi'_{\phi,i,1} + \sum_{\substack{i'=1\\ i'\neq i}}^{N} \xi_{i',i,1} \leq \sum_{\substack{i'=1\\ i'\neq i}}^{N} \xi_{i,i',1} + del_{i,0,1}$$

$$+(1 - \delta^1_{i,1} - \delta^2_{i,1} - \delta^3_{i,1}) \qquad \forall i. \tag{5.29}$$

$$\xi'_{\phi,i,1} + \sum_{\substack{i'=1\\ i'\neq i}}^{N} \xi_{i',i,1} \geq \sum_{\substack{i'=1\\ i'\neq i}}^{N} \xi_{i,i',1} + del_{i,0,1}$$

$$-(1 - \delta^1_{i,1} - \delta^2_{i,1} - \delta^3_{i,1}) \qquad \forall i. \tag{5.30}$$

$$\sum_{\substack{i'=1\\ i'\neq i}}^{N} (\xi'_{i',i,t} + \xi_{i',i,t}) \leq \sum_{\substack{i'=1\\ i'\neq i}}^{N} \xi_{i,i',t} + del_{i,0,t}$$

$$+(1 - \delta^1_{i,t} - \delta^2_{i,t} - \delta^3_{i,t}) \qquad \forall i \text{ and } t\text{=2,3,..,}T. \tag{5.31}$$

$$\sum_{\substack{i'=1\\ i'\neq i}}^{N} (\xi'_{i',i,t} + \xi_{i',i,t}) \geq \sum_{\substack{i'=1\\ i'\neq i}}^{N} \xi_{i,i',t} + del_{i,0,t}$$

$$-(1 - \delta^1_{i,t} - \delta^2_{i,t} - \delta^3_{i,t}) \qquad \forall i \text{ and } t\text{=2,3,..,}T. \tag{5.32}$$

/* Constraint (5.33) indicates that the setup indicator $\xi_{i',i,t}$ can exist only when the setup indicator $\xi'_{i',i,t}$ exists */

$$N \times \sum_{\substack{i'=1\\ i'\neq i}}^{N} \sum_{i=1}^{N} \xi'_{i',i,t} \geq \sum_{i'=1}^{N} \sum_{\substack{i=1\\ i'\neq i}}^{N} \xi_{i',i,t}, \qquad t\text{=2,3,..,}T. \tag{5.33}$$

/* Constraints (5.34)–(5.35) assign the value of the sequence-dependent setup time for product i, with its setup being initiated in period t, when a setup of product i takes place from product i' in period t */

$$ST'_{i,1} = (\xi'_{\phi,i,1} \times ST_{\phi,i}) + \sum_{\substack{i'=1 \\ i' \neq i}}^{N} (\xi_{i',i,1} \times ST_{i',i}) \quad \forall i. \tag{5.34}$$

$$ST'_{i,t} = \sum_{\substack{i'=1 \\ i' \neq i}}^{N} \left((\xi'_{i',i,t} + \xi_{i',i,t}) \times ST_{i',i} \right) \qquad \forall i \text{ and } t=2,3,..,T. \tag{5.35}$$

/* Constraint (5.36) is the sub-tour elimination constraint associated with $\xi_{i',i,t}$ */

$$U_{i',t} - U_{i,t} + N \times \xi_{i',i,t} \leq N - 1 \qquad \forall i, \forall i', i \neq i' \text{ and } \forall t. \tag{5.36}$$

/* Constraints (5.37)–(5.38) represent the conditions for setting up a product only once in period t */

$$\sum_{i=1}^{N} (\delta^2_{i,t} + \delta^3_{i,t}) \leq 1, \qquad t=1,2,..,T\text{-}1. \tag{5.37}$$

$$(\delta^1_{i,t} + \delta^2_{i,t} + \delta^3_{i,t}) \leq 1 \qquad \forall i \text{ and } t=1,2,..,T\text{-}1. \tag{5.38}$$

/* Constraints (5.39)–(5.47) capture a possible complete setup within period t, with the production starting in the same period t */

$$\Delta^1_{i,t,t} = \delta^1_{i,t} \qquad \forall i \text{ and } \forall t. \tag{5.39}$$

$$\Delta^1_{i,t,t'} \geq \Delta^1_{i,t,t'+1} \qquad \forall i, t=1,2,..,T\text{-}1 \text{ and } t'=t,t+1,..,T\text{-}1. \tag{5.40}$$

$$\sum_{i'=1}^{N} (\delta^1_{i',t'-1} + \delta^2_{i',t'-1} + \delta^3_{i',t'-1}) \leq N(1 - \Delta^1_{i,t,t'}) \quad \forall i, t=1,2,..,T\text{-}2 \text{ and } t'=t+2,t+3,..,T. \tag{5.41}$$

$$X^1_{i,t,t'} \geq \mathcal{E}_d - \frac{C_{t'}}{a_i}(1 - \Delta^1_{i,t,t'}) \qquad \forall i, \ \forall t \text{ and } t'=t,t+1,..,T. \tag{5.42}$$

$$a_i \, X^1_{i,t,t'} \leq C_{t'} \, \Delta^1_{i,t,t'} \qquad \forall i, \ \forall t \text{ and } t'=t,t+1,..,T. \tag{5.43}$$

$$s^1_{i,t} \leq ST'_{i,t} + M(1 - \delta^1_{i,t}) \qquad \forall i \text{ and } \forall t. \tag{5.44}$$

$$s^1_{i,t} \geq ST'_{i,t} - M(1 - \delta^1_{i,t}) \qquad \forall i \text{ and } \forall t. \tag{5.45}$$

$$s^1_{i,t} \leq M \, \delta^1_{i,t} \qquad \forall i \text{ and } \forall t. \tag{5.46}$$

$$s^1_{i,t} \leq (C_t - \mathcal{E}) + C_t(1 - \delta^1_{i,t}) \qquad \forall i \text{ and } \forall t. \tag{5.47}$$

/* Constraints (5.48)–(5.56) correspond to a possible end-of-period setup in period t, with the production starting in period $t+1$ */

$$\Delta^2_{i,t,t'} = \delta^2_{i,t} \quad \forall i,\ t=1,2,..,T\text{-}1 \text{ and } t'=t+1. \tag{5.48}$$

$$\Delta^2_{i,t,t'} \geq \Delta^2_{i,t,t'+1} \quad \forall i,\ t=1,2,..,T\text{-}2 \text{ and } t'=t+1,t+2,..,T\text{-}1. \tag{5.49}$$

$$\sum_{i'=1}^{N}(\delta^1_{i',t'-1}+\delta^2_{i',t'-1}+\delta^3_{i',t'-1}) \leq N(1-\Delta^2_{i,t,t'}) \quad \forall i,\ t=1,2,..,T\text{-}2 \text{ and } t'=t+2,t+3,..,T. \tag{5.50}$$

$$X^2_{i,t,t'} \geq \mathcal{E}_d - \frac{C_{t'}}{a_i}(1-\Delta^2_{i,t,t'}) \quad \forall i,\ t=1,2,..,T\text{-}1 \text{ and } t'=t+1,t+2,..,T. \tag{5.51}$$

$$a_i\, X^2_{i,t,t'} \leq C_{t'}\, \Delta^2_{i,t,t'} \quad \forall i,\ t=1,2,..,T\text{-}1 \text{ and } t'=t+1,t+2,..,T. \tag{5.52}$$

$$s^2_{i,t} \leq ST'_{i,t} + M(1-\delta^2_{i,t}) \quad \forall i \text{ and } t=1,2,..,T\text{-}1. \tag{5.53}$$

$$s^2_{i,t} \geq ST'_{i,t} - M(1-\delta^2_{i,t}) \quad \forall i \text{ and } t=1,2,..,T\text{-}1. \tag{5.54}$$

$$s^2_{i,t} \leq M\,\delta^2_{i,t} \quad \forall i \text{ and } t=1,2,..,T\text{-}1. \tag{5.55}$$

$$s^2_{i,t} \leq C_t + C_t(1-\delta^2_{i,t}) \quad \forall i\ t=1,2,..,T\text{-}1 \text{ and } t'=t+1. \tag{5.56}$$

/* Constraints (5.57)–(5.74) represent a possible setup crossover across a number of periods with the production starting in the same period where the setup ends */

$$\Omega^3_{i,t,t} \leq \delta^3_{i,t} \quad \forall i \text{ and } t=1,2,..,T\text{-}1. \tag{5.57}$$

$$\Omega^3_{i,t,t+1} = \Omega^3_{i,t,t} \quad \forall i \text{ and } t=1,2,..,T\text{-}1. \tag{5.58}$$

$$\Omega^3_{i,t,t'} \geq \Omega^3_{i,t,t'+1} \quad \forall i,\ t=1,2,..,T\text{-}1 \text{ and } t'=t,t+1,..,T\text{-}1. \tag{5.59}$$

$$\Omega^3_{i,t,t'} + \Delta^3_{i,t,t'} \geq \Delta^3_{i,t,t'+1} \quad \forall i,\ t=1,2,..,T\text{-}1 \text{ and } t'=t,t+1,..,T. \tag{5.60}$$

$$\Delta^3_{i,t,t'} \leq \Omega^3_{i,t,t} \quad \forall i,\ t=1,2,..,T\text{-}1 \text{ and } t'=t,t+1,..,T. \tag{5.61}$$

$$\Omega^3_{i,t,t'+1} + \Delta^3_{i,t,t'} \leq 1 \quad \forall i,\ t=1,2,..,T\text{-}1 \text{ and } t'=t+1,t+2,..,T. \tag{5.62}$$

$$\Omega^3_{i,t,t'} - \Omega^3_{i,t,t'+1} \leq \Delta^3_{i,t,t'} \quad \forall i,\ t=1,2,..,T\text{-}1 \text{ and } t'=t+1,t+2,..,T. \tag{5.63}$$

$$\sum_{i=1}^{N}\sum_{t''=1}^{t-1}\Omega^3_{i,t'',t} \leq 1, \quad t=2,3,..,T. \tag{5.64}$$

$$\sum_{i'=1}^{N}(\delta^1_{i',t'-1}+\delta^2_{i',t'-1}+\delta^3_{i',t'-1}) \leq N(1-\Delta^3_{i,t,t'}) \quad \forall i,\ t=1,2,..,T\text{-}2 \text{ and } t'=t+2,t+3,..,T. \tag{5.65}$$

$$\sum_{i'=1}^{N}(\delta^1_{i',t'-1}+\delta^2_{i',t'-1}+\delta^3_{i',t'-1}) \leq N(1-\Omega^3_{i,t,t'}) \quad \forall i,\ t=1,2,..,T\text{-}2 \text{ and } t'=t+2,t+3,..,T. \tag{5.66}$$

$$X^3_{i,t,t'} \geq \mathcal{E}_d - \frac{C_{t'}}{a_i}(1 - \Delta^3_{i,t,t'}) \qquad \forall i,\ t=1,2,..,T\text{-}1 \text{ and } t'=t+1,t+2,..,T. \tag{5.67}$$

$$a_i\, X^3_{i,t,t'} \leq C_{t'}\, \Delta^3_{i,t,t'} \qquad \forall i,\ t=1,2,..,T\text{-}1 \text{ and } t'=t+1,t+2,..,T. \tag{5.68}$$

$$s^3_{i,t,t'} \geq \mathcal{E} - C_{t'}(1 - \Omega^3_{i,t,t'}) \qquad \forall i,\ t=1,2,..,T\text{-}1 \text{ and } t'=t,t+1,..,T. \tag{5.69}$$

$$s^3_{i,t,t'} \leq C_{t'}\, \Omega^3_{i,t,t'} \qquad \forall i,\ t=1,2,..,T\text{-}1 \text{ and } t'=t,t+1,..,T. \tag{5.70}$$

$$\sum_{t'=t}^{T} s^3_{i,t,t'} \leq ST'_{i,t} + ((M+1) \times (1 - \Omega^3_{i,t,t})) \qquad \forall i \text{ and } t=1,2,..,T\text{-}1. \tag{5.71}$$

$$\sum_{t'=t}^{T} s^3_{i,t,t'} \geq ST'_{i,t} - ((M+1) \times (1 - \Omega^3_{i,t,t})) \qquad \forall i \text{ and } t=1,2,..,T\text{-}1. \tag{5.72}$$

$$s^3_{i,t,t'-1} \leq C_{t'-1} + ((C_{t'-1}+1) \times (1 - \Omega^3_{i,t,t'})) \qquad \forall i,\ t=1,2,..,T\text{-}2 \text{ and } t'=t+2,t+3,..,T. \tag{5.73}$$

$$s^3_{i,t,t'-1} \geq C_{t'-1} - ((C_{t'-1}+1) \times (1 - \Omega^3_{i,t,t'})) \qquad \forall i,\ t=1,2,..,T\text{-}2 \text{ and } t'=t+2,t+3,..,T. \tag{5.74}$$

/* Constraints (5.75)–(5.91) represent a possible setup crossover across a number of periods with the production starting in period $t'+1$, when the setup ends at the end-of-period t' */

$$\Omega^4_{i,t,t} \leq \delta^3_{i,t} \qquad \forall i \text{ and } t=1,2,..,T\text{-}2. \tag{5.75}$$

$$\Omega^4_{i,t,t+1} = \Omega^4_{i,t,t} \qquad \forall i \text{ and } t=1,2,..,T\text{-}2. \tag{5.76}$$

$$\Omega^4_{i,t,t'} \geq \Omega^4_{i,t,t'+1} \qquad \forall i,\ t=1,2,..,T\text{-}2 \text{ and } t'=t,t+1,..,T\text{-}1. \tag{5.77}$$

$$\Omega^4_{i,t,t'} + \Delta^4_{i,t,t'} \geq \Delta^4_{i,t,t'+1} \qquad \forall i,\ t=1,2,..,T\text{-}2 \text{ and } t'=t+1,t+2,..,T. \tag{5.78}$$

$$\Delta^4_{i,t,t'} \leq \Omega^4_{i,t,t} \qquad \forall i,\ t=1,2,..,T\text{-}2 \text{ and } t'=t,t+1,..,T. \tag{5.79}$$

$$\Omega^4_{i,t,t'} + \Delta^4_{i,t,t'} \leq 1 \qquad \forall i,\ t=1,2,..,T\text{-}2 \text{ and } t'=t+1,t+2,..,T\text{-}1. \tag{5.80}$$

$$\Omega^4_{i,t,t'} - \Omega^4_{i,t,t'+1} \leq \Delta^4_{i,t,t'+1} \qquad \forall i,\ t=1,2,..,T\text{-}2 \text{ and } t'=t+1,t+2,..,T\text{-}1. \tag{5.81}$$

$$\sum_{i=1}^{N} \sum_{t''=1}^{t-1} \Omega^4_{i,t'',t} \leq 1, \qquad t=2,3,..,T\text{-}1. \tag{5.82}$$

$$\sum_{i'=1}^{N} (\delta^1_{i',t'-1} + \delta^2_{i',t'-1} + \delta^3_{i',t'-1}) \leq N(1 - \Delta^4_{i,t,t'}) \qquad \forall i,\ t=1,2,..,T\text{-}2 \text{ and } t'=t+2,t+3,..,T. \tag{5.83}$$

$$\sum_{i'=1}^{N} (\delta^1_{i',t'-1} + \delta^2_{i',t'-1} + \delta^3_{i',t'-1}) \leq N(1 - \Omega^4_{i,t,t'}) \qquad \forall i,\ t=1,2,..,T\text{-}2 \text{ and } t'=t+2,t+3,..,T. \tag{5.84}$$

$$X^4_{i,t,t'} \geq \mathcal{E}_d - \frac{C_{t'}}{a_i}(1 - \Delta^4_{i,t,t'}) \qquad \forall i,\ t=1,2,..,T\text{-}2 \text{ and } t'=t+2,t+3,..,T. \tag{5.85}$$

$$a_i\, X^4_{i,t,t'} \leq C_{t'}\, \Delta^4_{i,t,t'} \qquad \forall i,\ t=1,2,..,T\text{-}2 \text{ and } t'=t+2,t+3,..,T. \tag{5.86}$$

$$s^4_{i,t,t} \geq \mathcal{E} - C_t(1 - \Omega^4_{i,t,t}) \qquad \forall i \text{ and } t=1,2,..,T\text{-}2. \tag{5.87}$$

$$s^4_{i,t,t} \leq C_t\, \Omega^4_{i,t,t} \qquad \forall i \text{ and } t=1,2,..,T\text{-}2. \tag{5.88}$$

$$\textstyle\sum_{t'=t}^{T-1} s^4_{i,t,t'} \leq ST'_{i,t} + ((M+1) \times (1 - \Omega^4_{i,t,t})) \qquad \forall i \text{ and } t=1,2,..,T\text{-}2. \tag{5.89}$$

$$\textstyle\sum_{t'=t}^{T-1} s^4_{i,t,t'} \geq ST'_{i,t} - ((M+1) \times (1 - \Omega^4_{i,t,t})) \qquad \forall i \text{ and } t=1,2,..,T\text{-}2. \tag{5.90}$$

$$s^4_{i,t,t'} = C_{t'}\, \Omega^4_{i,t,t'} \qquad \forall i,\ t=1,2,..,T\text{-}2 \text{ and } t'=t+1,t+2,..,T\text{-}1. \tag{5.91}$$

/* Constraints (5.92) and (5.93) indicate that either $\Omega^3_{i,t,t}$ or $\Omega^4_{i,t,t}$ can only exist in period t when $\delta^3_{i,t} = 1$ */

$$\Omega^3_{i,t,t} + \Omega^4_{i,t,t} = \delta^3_{i,t} \qquad \forall i \text{ and } t=1,2,..,T\text{-}1. \tag{5.92}$$

$$\Omega^3_{i,t,t'} + \Omega^4_{i,t,t'} \leq \delta^3_{i,t} \qquad \forall i,\ t=1,2,..,T\text{-}1 \text{ and } t'=t,t+1,..,T. \tag{5.93}$$

/* Constraints (5.94) and (5.95) represent the condition that only one type of production carryover can take place for at most one product in a period */

$$\textstyle\sum_{i=1}^{N}(\Delta^1_{i,1,2} + \Delta^2_{i,1,2} + \Delta^3_{i,1,2}) \leq 1. \tag{5.94}$$

$$\textstyle\sum_{i=1}^{N}\sum_{t''=1}^{t-1}(\Delta^1_{i,t'',t} + \Delta^2_{i,t'',t} + \Delta^3_{i,t'',t}) + \sum_{i=1}^{N}\sum_{t''=1}^{t-2}\Delta^4_{i,t'',t} \leq 1, \qquad t=3,4,..,T. \tag{5.95}$$

/* Constraint (5.96) indicates that any production carryover and setup crossover of product i cannot take place from period t to period t' if a setup of product i' succeeds the setup of product i in period t */

$$\textstyle\sum_{t'=t}^{T-1} \Delta^1_{i,t,t'+1} \leq T \times (1 - \sum_{i'=1, i'\neq i}^{N} \xi_{i,i',t}) \qquad \forall i \text{ and } t=1,2,..,T\text{-}1. \tag{5.96}$$

/* Constraints (5.97)–(5.99) show that if the setup of at least 1 product is initiated in period t (indicated by $delta_t = 1$), then it means that no carryover of either setup or production of any product, with its setup having been initiated in any period up to $t-1$, can be carried over to period $t+1$ and later. This is so because if any product's setup is initiated in period t, then either the production or setup carryover with the setup initiated up to period $t-1$ and carried over to period $t+1$ and later is not feasible. */

$$\sum_{i=1}^{N}\sum_{t''=1}^{t-1}\sum_{t'''=t+1}^{T}(\Delta^1_{i,t'',t'''}+\Delta^2_{i,t'',t'''}$$

$$+\Delta^3_{i,t'',t'''}+\Delta^4_{i,t'',t'''}$$

$$+\Omega^3_{i,t'',t'''}+\Omega^4_{i,t'',t'''})\leq 6\times N\times T^2(1-delta_t),\quad t\text{=2,3,..,}T\text{-1.} \tag{5.97}$$

$$\sum_{i=1}^{N}(\delta^1_{i,t}+\delta^2_{i,t}+\delta^3_{i,t})\leq (N+1)\times delta_t,\quad t\text{=2,3,..,}T\text{-1.} \tag{5.98}$$

$$delta_t\leq \sum_{i=1}^{N}(\delta^1_{i,t}+\delta^2_{i,t}+\delta^3_{i,t}),\quad t\text{=2,3,..,}T\text{-1.} \tag{5.99}$$

/* Conditions (5.100) and (5.101) indicate that a production carryover for any product i can take place in period $t+2\leq t'\leq T$, only if the sum of the production times and setup times of all products in period $t'-1$ is equal to the capacity of period $t'-1$ */

$$\sum_{i=1}^{N}(a_i\,X^1_{i,t,t'-1}+a_i\,X^2_{i,t,t'-1}+a_i\,X^3_{i,t,t'-1}$$

$$+a_i\,X^4_{i,t,t'-1}+s^3_{i,t,t'-1}+s^4_{i,t,t'-1})\leq C_{t'-1}$$

$$+((C_{t'-1}+1)\times(1-\sum_{i=1}^{N}(\Delta^1_{i,t,t'}+\Delta^2_{i,t,t'}$$
$$+\Delta^3_{i,t,t'}+\Delta^4_{i,t,t'}))),\quad t\text{=1,2,..,}T\text{-2 and }t'\text{=}t\text{+2,}t\text{+3,..,}T. \tag{5.100}$$

$$\sum_{i=1}^{N}(a_i\,X^1_{i,t,t'-1}+a_i\,X^2_{i,t,t'-1}+a_i\,X^3_{i,t,t'-1}$$

$$+a_i\,X^4_{i,t,t'-1}+s^3_{i,t,t'-1}+s^4_{i,t,t'-1})\geq C_{t'-1}$$

$$-((C_{t'-1}+1)\times(1-\sum_{i=1}^{N}(\Delta^1_{i,t,t'}+\Delta^2_{i,t,t'}$$

$$+\Delta^3_{i,t,t'}+\Delta^4_{i,t,t'}))),\quad t\text{=1,2,..,}T\text{-2 and }t'\text{=}t\text{+2,}t\text{+3,..,}T. \tag{5.101}$$

/* Constraints (5.102)–(5.104) represent the feasibility of production and setup with respect to capacity availability */

$$\sum_{i=1}^{N}(a_i\,X^1_{i,1,1}+s^1_{i,1}+s^2_{i,1}+s^3_{i,1,1}+s^4_{i,1,1})\leq C_1. \tag{5.102}$$

$$\sum_{i=1}^{N}(s^1_{i,t}+s^2_{i,t}+\sum_{t''=1}^{t}s^3_{i,t'',t}+\sum_{t''=1}^{t}s^4_{i,t'',t})$$

$$+\sum_{i=1}^{N}\sum_{t''=1}^{t}a_i\,X^1_{i,t'',t}$$

$$+\sum_{i=1}^{N}\sum_{t''=1}^{t-1}(a_i\,X^2_{i,t'',t}+a_i\,X^3_{i,t'',t}$$

$$+a_i\,X^4_{i,t'',t})\leq C_t,\quad t\text{=2,3,..,}T\text{-1.} \tag{5.103}$$

$$\sum_{i=1}^{N} s_{i,T}^{1} + \sum_{i=1}^{N} \sum_{t''=1}^{T-1} s_{i,t'',T}^{3}$$

$$+ \sum_{i=1}^{N} \sum_{t''=1}^{T} a_i\, X_{i,t'',T}^{1}$$

$$+ \sum_{i=1}^{N} \sum_{t''=1}^{T-1} (a_i\, X_{i,t'',T}^{2} + a_i\, X_{i,t'',T}^{3}$$

$$+ a_i\, X_{i,t'',T}^{4}) \le C_T. \tag{5.104}$$

/* Constraints (5.105)–(5.106) represent the inventory balance constraints */

$$I_{i,1} - B_{i,1} = X_{i,1,1}^{1} - d_{i,1} \qquad \forall i. \tag{5.105}$$

$$I_{i,t} - B_{i,t} = \sum_{t'''=1}^{t} \sum_{t''=1}^{t'''} X_{i,t'',t'''}^{1}$$

$$+ \sum_{t'''=2}^{t} \sum_{t''=1}^{t'''-1} (X_{i,t'',t'''}^{2} + X_{i,t'',t'''}^{3}$$

$$+ X_{i,t'',t'''}^{4}) - \sum_{t'''=1}^{t} d_{i,t'''} \qquad \forall i \text{ and } t=2,3,..,T. \tag{5.106}$$

/* Boundary conditions */

$$\delta_{i,T}^{2} = \delta_{i,T}^{3} = \Omega_{i,T,T}^{3} = I_{i,T} = B_{i,T}$$

$$= \Omega_{i,T-1,T-1}^{4} = \Omega_{i,T-1,T}^{4} = \Delta_{i,T-1,T}^{4}$$

$$= \Delta_{i,T-1,T+1}^{4} = \delta_{i,0}^{1} = \delta_{i,0}^{2} = \delta_{i,0}^{3} = 0 \qquad \forall i. \tag{5.107}$$

$$\xi'_{i,i,t} = \xi_{i,i,t} = \Delta_{i,t,t}^{2} = X_{i,t,t}^{2}$$

$$= \Delta_{i,t,t}^{3} = \Delta_{i,t,T+1}^{3} = X_{i,t,t}^{3}$$

$$= X_{i,t,T+1}^{3} = \Omega_{i,t,T+1}^{3} = \Delta_{i,t,t}^{4} = \Delta_{i,t,t+1}^{4}$$

$$= \Delta_{i,t,T+1}^{4} = X_{i,t,T+1}^{4} = X_{i,t,t}^{4} = X_{i,t,t+1}^{4}$$

$$= s_{i,t,T}^{4} = \Omega_{i,t,T}^{4} = 0 \qquad \forall i \text{ and } \forall t. \tag{5.108}$$

$\delta_{i,t}^{1}$, $\delta_{i,t}^{2}$, $\delta_{i,t}^{3}$, $\Delta_{i,t,t'}^{1}$, $\Delta_{i,t,t'}^{2}$, $\Delta_{i,t,t'}^{3}$, $\Delta_{i,t,t'}^{4}$, $\Omega_{i,t,t'}^{3}$, $\Omega_{i,t,t'}^{4}$, $delta_t$, $del_{\phi,i,1}$, $del_{0,i,t}$, $del_{i,0,t}$, $\xi'_{\phi,i,1}$, $\xi'_{i,i',t}$ and $\xi_{i',i,t}$ are binary for all i, t and t' (with $t \le t' \le T$ or $t+1 \le t' \le T$, as appropriate to the respective binary variable), and all other variables are ≥ 0.

The objective function shown in Eq. (5.1) denotes the minimization of sequence-dependent setup costs, holding costs and backorder costs of all products across all time periods. Constraints (5.2)–(5.5) indicate that only one product should be tracked (concerning its setup/production) with respect to the beginning and end of every period t. Constraint (5.6) indicates that the product tracked last (concerning its setup and production) with respect to period $t-1$ is equal to the product tracked last with respect to the beginning of period t. Constraint (5.7) helps to track the product setup with respect to the beginning of period 1. Constraints (5.8)–(5.16) help to track the product that has been set up last with respect to the end of period t. Constraints (5.17)–(5.19) assign the product setup with respect to the beginning of every period t to the indicator variable. Constraint (5.20) indicates that at least one product should be set up as the first setup in period 1. Constraint (5.21) indicates that the binary variable takes the value 1 if at least one product i' is set up as the first setup in period t. Constraints (5.22)–(5.24) assign the value 1 to the indicator variable when product i is tracked for its setup with respect to the end of every period t. Constraints (5.25)–(5.28) link the setup of product i denoted by $\delta^1_{i,t}$, $\delta^2_{i,t}$ and $\delta^3_{i,t}$ with the indicator variable which assigns the setup of product i initiated in period t preceded by product i'. Constraints (5.29)–(5.32) represent the flow balance constraints with respect to setup carryover tracking and the current period's setup tracking. Constraint (5.33) indicates that the setup indicator $\xi_{i',i,t}$ can exist only when the setup indicator $\xi'_{i',i,t}$ exists. Constraints (5.34)–(5.35) assign the value of the sequence-dependent setup time for product i, with its setup being initiated in period t, when a setup of product i takes place from product i' in period t. Constraint (5.36) is the sub-tour elimination constraint associated with $\xi_{i',i,t}$. Constraints (5.37)–(5.38) represent the conditions for setting up a product only once in period t. Constraints (5.39)–(5.47) capture a possible complete setup within period t, with the production starting in the same period t. Constraints (5.48)–(5.56) correspond to a possible end-of-period setup in period t, with the production starting in period $t+1$. Constraints (5.57)–(5.74) represent a possible setup crossover across a number of periods with the production starting in the same period where the setup ends. Constraints (5.75)–(5.91) represent a possible setup crossover across a number of periods with the production starting in period $t'+1$, when the setup ends at the end-of-period t'. Constraints (5.92) and (5.93) indicate that either $\Omega^3_{i,t,t}$ or $\Omega^4_{i,t,t}$ can only exist in period t when $\delta^3_{i,t} = 1$. Constraints (5.94) and (5.95) represent the condition that only one type of production carryover can take place for at most one product in a period. Constraint (5.96) indicates that any production carryover and setup crossover of product i cannot take place from period t to period t' if a setup of product i' succeeds the setup of product i in period t. Constraints (5.97)–(5.99) show that if the setup of at least 1 product is initiated in period t (indicated by $delta_t$=1), then it means that no carryover of either setup or production of any product, with its setup having been initiated in any period up to $t-1$, can be carried over to period $t+1$ and later. This is so because if any product's setup is initiated in period t, then either the production or setup carryover with the setup initiated up to period $t-1$ and carried over to period $t+1$ and later is not feasible. Conditions (5.100) and (5.101) indicate that a production carryover for any product i can take place in period $t+2 \leq t' \leq T$, only if the sum of the production times and setup times of all products in period t'-1 is equal to the capacity of period t'-1. Constraints (5.102)–(5.104) represent the feasibility of production and setup with respect to capacity availability. Constraints (5.105)–(5.106) represent the inventory balance constraints. Constraints (5.107) and (5.108) are the boundary conditions.

Also in Eq. (5.107), $B_{i,T} = 0 \;\forall i$ exists because production is not allowed to be carried over to period $T+1$ (i.e. here the demand for all products should be satisfied within period T).

5.4 Mathematical Model 2 (MM2:CLSP-SD-PCSC)

5.4.1 Parameters/Indices

N	number of products
T	number of time periods
t	a given time period
i	product
$SC_{\phi,i}$	sequence-dependent setup cost incurred in the machine for the first product i setup in period 1
$SC_{i',i}$	sequence-dependent setup cost incurred when the machine is set up from product i' to product i
b_i	backorder cost per period per unit of product i
h_i	holding cost per period per unit of product i
$ST_{\phi,i}$	sequence-dependent setup time for the first product i setup in the machine in period 1
$ST_{i',i}$	sequence-dependent setup time when the machine is set up from product i' to product i
a_i	number of time units required for producing one unit of product i
C_t	capacity of the machine in period t (in time units)
$d_{i,t}$	demand for product i in period t
$\mathcal{E}$	smallest unit of time
$\mathcal{E}_d$	unit of smallest quantity of production

5.4.2 Decision Variables

Variable	Description
$\delta^1_{0,i,t}$	An indicator (binary) variable that takes value 1 if a complete setup is done for product i in the beginning of period $t = 1$ (as the first setup) with the production starting in period 1; 0 otherwise.
$\delta^1_{i',i,t}$	An indicator (binary) variable that takes value 1 if a complete setup is done for product i (after setting up product i') in the beginning of period t with the production starting in period t; 0 otherwise.
$\Delta^1_{i,t,t'}$	An indicator (binary) variable that takes value 1: it corresponds to the production carryover from period t to period t' ($t \leq t' \leq T$), due to the setup of product i started and finished in the beginning of period t, with no intermittent setup of any other product; 0 otherwise.
$\delta^2_{i',i,t}$	An indicator (binary) variable that takes value 1 if a setup is done for product i (after setting up product i') anywhere in the middle of period t except in the beginning or the end of period t followed by its production starting in period t; 0 otherwise.

Variable	Description
$\Delta^2_{i,t,t'}$	An indicator (binary) variable that takes value 1: it corresponds to the production carryover from period t to period t' ($t \leq t' \leq T$), due to the setup of product i started and finished anywhere in the middle of period t, with no intermittent setup of any other product; 0 otherwise.
$\delta^3_{i',i,t}$	An indicator (binary) variable that takes value 1 if a setup is done for product i (after setting up product i') at the end-of-period t followed by its production starting in period $t+1$; 0 otherwise.
$\Delta^3_{i,t,t'}$	An indicator (binary) variable that takes value 1: it corresponds to the production carryover from period t' to period $t'+1$ ($t+1 \leq t' \leq T$), due to the setup of product i started and completed exactly at the end-of period t, with no intermittent setup of any other product; 0 otherwise.
$\delta^4_{0,i,t}$	An indicator (binary) variable that takes value 1 if a complete setup is done for product i in period $t = 1$ (as the first setup) with the setup started exactly in the beginning of period t, and completed exactly at the end of period t, and with the production starting in period $t+1$. The setup time of the product setup using this indicator binary variable should be equal to the capacity of period 1; 0 otherwise.
$\delta^4_{i',i,t}$	An indicator (binary) variable that takes value 1 if a complete setup is done for product i in period $t = 2, 3, .., T-1$ with the setup started exactly in the beginning of period t, and completed exactly at the end of period t, and with the production starting in period $t+1$. The setup time of the product setup using this indicator binary variable should be equal to the capacity of period t; 0 otherwise.
$\Delta^4_{i,t,t'}$	An indicator (binary) variable that takes value 1: it corresponds to the production carryover from period t' to period $t'+1$ ($t+1 \leq t' \leq T$), due to the complete setup of product i in period t (with the setup started exactly in the beginning of period t, and completed exactly at the end of period t), and with no intermittent setup of any other product; 0 otherwise.
$\delta^5_{i',i,t,t'}$	An indicator (binary) variable that takes value 1 if the setup of product i is commenced in the beginning of period t and is completed during the middle of some period t' but not at the end of period t' ($t+1 \leq t' \leq T$), followed by its production starting in period t'; 0 otherwise.

Variable	Description
$\Delta^5_{i,t,t',t''}$	An indicator (binary) variable that takes value 1: it corresponds to the production in period t'' ($t' \leq t'' \leq T$), due to the setup of product i initiated in the beginning of period t and completed in the middle of period t' but not at the end of period t' ($t+1 \leq t' \leq T$), and with no setup of any product during the intermittent periods from period t' to period $t'' > t'$. *Note:* This variable corresponds to (i.e. indicates) the production carryover through periods t' and $t'' > t'$, after the completion of setup in the middle of period t', with the initiation of setup in the beginning of an earlier period t; 0 otherwise.
$\Omega^5_{i,t,t',t''}$	An indicator (binary) variable that takes value 1 if the setup of product i is initiated in the beginning of period t and gets completed in the middle of a later period t' ($t+1 \leq t' \leq T$) but not exactly at the end of that period, with the setup of that product being present in period t'' ($t \leq t'' \leq t'$); 0 otherwise.
$\delta^6_{i',i,t,t'}$	An indicator (binary) variable that takes value 1 if the setup of product i is commenced in the beginning of period t and is completed exactly at the end of period t' ($t+1 \leq t' \leq T-1$), followed by its production starting in period $t'+1$; 0 otherwise.
$\Delta^6_{i,t,t',t''}$	An indicator (binary) variable that takes value 1: it corresponds to the production in period t'' ($t'+1 \leq t'' \leq T$), due to the setup of product i initiated in the beginning of period t and completed exactly at the end of period t' ($t+1 \leq t' \leq T-1$), and with no setup of any product during the intermittent periods from period $t'+1$ to period $t'' > t'+1$. *Note:* This variable corresponds to (i.e. indicates) the production carryover through periods $t'+1$ and $t'' > t'+1$, after the completion of setup exactly at the end of period t', with the initiation of setup in the beginning of an earlier period t; 0 otherwise.
$\Omega^6_{i,t,t',t''}$	An indicator (binary) variable that takes value 1 if the setup of product i is initiated in the beginning of period t and gets completed exactly at the end of period t' ($t+1 \leq t' \leq T-1$), with the setup of that product being present in period t'' ($t \leq t'' \leq t'$); 0 otherwise.
$\delta^7_{i',i,t,t'}$	An indicator (binary) variable that takes value 1 if the setup of product i is commenced in the middle of period t and is completed during the middle of some period t' but not exactly at the end of period t' ($t+1 \leq t' \leq T$), followed by its production starting in period t'; 0 otherwise.

Variable	Description
$\Delta^7_{i,t,t',t''}$	An indicator (binary) variable that takes value 1: it corresponds to the production in period t'' ($t' \leq t'' \leq T$), due to the setup of product i initiated in the middle of period t and completed in the middle of period t' but not exactly at the end of period t' ($t+1 \leq t' \leq T$), and with no setup of any product during the intermittent periods from period t' to period $t'' > t'$. *Note:* This variable corresponds to (i.e. indicates) the production carryover through periods t' and $t'' > t'$, after the completion of setup in the middle of period t', with the initiation of setup in the middle of an earlier period t; 0 otherwise.
$\Omega^7_{i,t,t',t''}$	An indicator (binary) variable that takes value 1 if the setup of product i is initiated in the middle of period t and gets completed in the middle of a later period t' ($t+1 \leq t' \leq T$) but not exactly at the end of that period, with the setup of that product being present in period t'' ($t \leq t'' \leq t'$); 0 otherwise.
$\delta^8_{i',i,t,t'}$	An indicator (binary) variable that takes value 1 if the setup of product i is commenced in the middle of period t and is completed exactly at the end of period t' ($t+1 \leq t' \leq T-1$), followed by its production starting in period $t'+1$; 0 otherwise.
$\Delta^8_{i,t,t',t''}$	An indicator (binary) variable that takes value 1: it corresponds to the production in period t'' ($t'+1 \leq t'' \leq T$), due to the setup of product i initiated in the middle of period t and completed exactly at the end of period t' ($t+1 \leq t' \leq T-1$), and with no setup of any product during the intermittent periods from period $t'+1$ to period $t'' > t'+1$. *Note:* This variable corresponds to (i.e. indicates) the production carryover through periods $t'+1$ and $t'' > t'+1$, after the completion of setup at the end of period t', with the initiation of setup in the middle of an earlier period t; 0 otherwise.
$\Omega^8_{i,t,t',t''}$	An indicator (binary) variable that takes value 1 if the setup of product i is initiated in the middle of period t and gets completed in the middle of a later period t' ($t+1 \leq t' \leq T-1$), with the setup of that product being present in period t'' ($t \leq t'' \leq t'$); 0 otherwise.
$I_{i,t}$	Inventory of product i at the end of period t.
$B_{i,t}$	Backorder quantity of product i at the end of period t.
$s^1_{i,t}$	Setup time of product i that takes the value of $ST_{0,i}$ when associated with $\delta^1_{0,i,t}$, in period $t=1$; and setup time of product i that takes the value of $ST_{i',i}$ when associated with $\delta^1_{i',i,t}$, in periods $t = 2, 3, .., T$.
$s^2_{i,t}$	Setup time of product i in period t that takes the value of $ST_{i',i}$, and associated with $\delta^2_{i',i,t}$.

Variable	Description
$s^3_{i,t}$	Setup time of product i in period t that takes the value of $ST_{i',i}$, and associated with $\delta^3_{i',i,t}$.
$s^4_{i,t}$	Setup time of product i that takes the value of $ST_{0,i}$ when associated with $\delta^4_{0,i,t}$, in period $t = 1$; and setup time of product i that takes the value of $ST_{i',i}$ when associated with $\delta^4_{i',i,t}$, in periods $t = 2, 3, .., T - 1$.
$s^5_{i,t,t',t''}$	Setup time of product i in period t'' ($t \leq t'' \leq t'$), when its setup has started in the beginning of period t and completed in the middle of period t' ($t+1 \leq t' \leq T$), but not completed exactly at the end of period t', and associated with $\delta^5_{i',i,t,t'}$; *Note:* $\sum_{t''=t}^{t'} s^5_{i,t,t',t''} = ST_{i',i}$.
$s^6_{i,t,t',t''}$	Setup time of product i in period t'' ($t \leq t'' \leq t'$), when its setup has started in the beginning of period t and completed exactly at the end of period t' ($t+1 \leq t' \leq T-1$), and associated with $\delta^6_{i',i,t,t'}$; *Note:* $\sum_{t''=t}^{t'} s^6_{i,t,t',t''} = ST_{i',i}$.
$s^7_{i,t,t',t''}$	Setup time of product i in period t'' ($t \leq t'' \leq t'$), when its setup has started in the middle of period t and completed in the middle of period t' ($t+1 \leq t' \leq T$), and associated with $\delta^7_{i',i,t,t'}$; *Note:* $\sum_{t''=t}^{t'} s^7_{i,t,t',t''} = ST_{i',i}$.
$s^8_{i,t,t',t''}$	Setup time of product i in period t'' ($t \leq t'' \leq t'$), when its setup has started in the middle of period t and completed exactly at the end of period t' ($t+1 \leq t' \leq T-1$), and associated with $\delta^8_{i',i,t,t'}$; *Note:* $\sum_{t''=t}^{t'} s^8_{i,t,t',t''} = ST_{i',i}$.
$X^1_{i,t,t'}$	Production quantity of product i in period t' (due to setup started in the beginning of period t, and ended before the end of period t), with $1 \leq t \leq T$ and $t \leq t' \leq T$, and associated with $\Delta^1_{i,t,t'}$.
$X^2_{i,t,t'}$	Production quantity of product i in period t' (due to its setup initiated in the middle of period t but not at the beginning or end of period t, and completed before thc end of period t), with $1 \leq t \leq T$ and $t \leq t' \leq T$, and associated with $\Delta^2_{i,t,t'}$.
$X^3_{i,t,t'}$	Production quantity of product i in period t' (due to its setup started and completed exactly at the end of period t), with $1 \leq t \leq T-1$ and $t+1 \leq t' \leq T$, and associated with $\Delta^3_{i,t,t'}$.
$X^4_{i,t,t'}$	Production quantity of product i in period t' (due to its setup initiated at the beginning of period t and completed exactly at the end of period t, with the setup time of product i occupying the entire capacity of period t), with $1 \leq t \leq T-1$ and $t+1 \leq t' \leq T$, and associated with $\Delta^4_{i,t,t'}$.
$X^5_{i,t,t',t''}$	Production quantity of product i in period t'' (due to its setup started in the beginning of period t, and completed in the middle of period t' but not completed exactly at the end of period t'), with $1 \leq t \leq T-1$, $t+1 \leq t' \leq T$ and $t+1 \leq t'' \leq T$, and $t'' \geq t'$, and associated with $\Delta^5_{i,t,t',t''}$.
$X^6_{i,t,t',t''}$	Production quantity of product i in period t'' (due to its setup started in the beginning of period t and completed exactly at the end of period t'), with $1 \leq t \leq T-2$, $t+1 \leq t' \leq T-1$ and $t+2 \leq t'' \leq T$, and $t'' > t'$, and associated with $\Delta^6_{i,t,t',t''}$.

Variable	Description
$X^7_{i,t,t',t''}$	Production quantity of product i in period t'' (due to its setup started in the middle of period t, and completed in the middle of period t' but not completed exactly at the end of period t'), with $1 \leq t \leq T-1$, $t+1 \leq t' \leq T$ and $t+1 \leq t'' \leq T$, and $t'' \geq t'$, and associated with $\Delta^7_{i,t,t',t''}$.
$X^8_{i,t,t',t''}$	Production quantity of product i in period t'' (due to its setup started in the middle of period t and completed exactly at the end of period t'), with $1 \leq t \leq T-2$, $t+1 \leq t' \leq T-1$ and $t+2 \leq t'' \leq T$, and $t'' > t'$, and associated with $\Delta^8_{i,t,t',t''}$.
$delta_t$	An indicator (binary) variable which takes the value 1, if the setup of at least one of any product is initiated in period t (indicated by $delta_t = 1$). It also means that no carryover of either setup or production of any product, with its setup having been initiated in any period up to period $t-1$, can be carried over to period $t+1$ and later. This is so because if any product's setup is initiated in period t, then either the production or setup carryover with the setup initiated up to period $t-1$ which can carried over to period $t+1$ and later is not feasible.
$U_{i,t}$	An auxiliary variable that assigns product i in period t; it helps to avoid subtours.
$\alpha_{i,t}$	An indicator (binary) variable which tracks product i that has been setup last, with respect to the beginning of time period t.
$\beta_{i,t}$	An indicator (binary) variable which tracks product i that has been setup last, with respect to the end of time period t.

5.4.3 Mathematical Model 2 (MM2:CLSP-SD-PCSC)

In this section, the mathematical model (MM2:CLSP-SD-PCSC) is presented, with an objective of minimizing the sequence-dependent setup costs, holding costs and backorder costs of all products across all time periods. The proposed mathematical model assumes sequence-dependent setup costs and setup times similar to MM1:CLSP-SD-PCSC. The triangular inequality with respect to setup times of products i, i' and i'' is also assumed (i.e. $ST_{i,i'} + ST_{i',i''} \geq ST_{i,i''}$ and $ST_{\phi,i} + ST_{\phi,i'} \geq ST_{\phi,i''}$), and so is the case with respect to setup costs of products i, i' and i''. This mathematical model helps to address production situations in process industries. They address situations such as the presence of long setup times (sequence dependent), production starting immediately after the product's setup and uninterrupted production carryover across periods. In this mathematical model, eight binary variables are used to track the various types of setup that can be performed in a period. Out of these eight types of setup, four types of setup are performed for the products having a small setup time (i.e. $ST_i \leq C_t \, \forall i$), they are a complete setup done (started and finished) in the beginning of period t (but not in the middle or exactly at the end of period t) followed by its production in period t; a complete setup done for product i (after setting up product i') anywhere in the middle of period t except in the beginning or the end of period t followed by its production starting in period t; a setup done for product i (after setting up product i') at the end-of-period t followed by its production starting in period $t+1$; and a setup started exactly in the beginning of period t, and completed exactly at the end of period t, with the production starting in period $t+1$. The other four types of setup are performed in the presence of long setup times (i.e. $ST_i \geq C_t \, \forall i$), they are a setup of product i commenced in the beginning of period t and completed during the middle of some period t' but not at the end of period t' followed by its production starting in period t'; a setup of product i commenced in the beginning of period t and completed exactly at the end of period t' followed by its production starting in period $t'+1$; the setup of product i commenced in the middle of period t and is completed during the middle of some period t' but not exactly at the end of period t' followed by its production starting in period t' (Note: this type of setup can also take place when the setup time of a product is less than the capacity of a period); and a setup of product i commenced in the middle of period t and completed exactly at the end of period t' followed by its production starting in period $t'+1$. These binary variables which indicate a product's setup are in turn linked with production carryover indicator variables that help to track the time period in which the product's setup is initiated and completed, and the time period in which the corresponding production is carried out. Production variables corresponding to these production carryover indicator variables are also present which help to determine the production time (here, the production quantity is measured in time units). There are variables which help to determine the setup time of a product in a period. When the setup of product i crosses over a number of periods, the time taken to setup product i is also split across these periods. There are also constraints to ensure that production starts immediately after setup and uninterrupted production takes place across periods. Through this mathematical model, it is ensured that the demand for all products is satisfied across the entire time horizon with the condition that the production time and the setup time of the products setup in a period do not exceed the capacity limitations (measured in time units) of that period.

Objective Function:

$$\text{Min } Z = \sum_{i=1}^{N} SC_{0,i}\,\delta^1_{0,i,1} + \sum_{\substack{i'=1\\ i'\neq i}}^{N}\sum_{i=1}^{N}\sum_{t=2}^{T} SC_{i',i}\,\delta^1_{i',i,t} + \sum_{\substack{i'=1\\ i'\neq i}}^{N}\sum_{i=1}^{N}\sum_{t=1}^{T} SC_{i',i}\,\delta^2_{i',i,t} + \tag{5.109}$$

$$\sum_{\substack{i'=1\\ i'\neq i}}^{N}\sum_{i=1}^{N}\sum_{t=1}^{T-1} SC_{i',i}\,\delta^3_{i',i,t} + \sum_{i=1}^{N} SC_{0,i}\,\delta^4_{0,i,1} + \sum_{\substack{i'=1\\ i'\neq i}}^{N}\sum_{i=1}^{N}\sum_{t=2}^{T-1} SC_{i',i}\,\delta^4_{i',i,t} +$$

$$\sum_{\substack{i'=1\\ i'\neq i}}^{N}\sum_{i=1}^{N}\sum_{t'=t+1}^{T}\sum_{t=1}^{T-1} SC_{i',i}\,\delta^5_{i',i,t,t'} + \sum_{\substack{i'=1\\ i'\neq i}}^{N}\sum_{i=1}^{N}\sum_{t'=t+1}^{T-1}\sum_{t=1}^{T-2} SC_{i',i}\,\delta^6_{i',i,t,t'} +$$

$$\sum_{\substack{i'=1\\ i'\neq i}}^{N}\sum_{i=1}^{N}\sum_{t'=t+1}^{T}\sum_{t=1}^{T-1} SC_{i',i}\,\delta^7_{i',i,t,t'} + \sum_{\substack{i'=1\\ i'\neq i}}^{N}\sum_{i=1}^{N}\sum_{t'=t+1}^{T-1}\sum_{t=1}^{T-2} SC_{i',i}\,\delta^8_{i',i,t,t'} + \sum_{i=1}^{N}\sum_{t=1}^{T} h_i\,I_{i,t} +$$

$$\sum_{i=1}^{N}\sum_{t=1}^{T} b_i\,B_{i,t}$$

subject to the following:

/* Constraints (5.110)–(5.118) capture a possible complete setup in the beginning of period t, with the production starting in period t. The setup time of the product being set up (using these constraints) should be less than the capacity of that period */

$$\Delta^1_{i,t,t} = \delta^1_{0,i,t} \qquad \forall i \text{ and } t\text{=1}. \qquad (5.110)$$

$$\Delta^1_{i,t,t} = \sum_{i'=1}^{N} \delta^1_{i',i,t} \qquad \forall i \text{ and } t\text{=2,3,..,}T. \qquad (5.111)$$

$$\Delta^1_{i,t,t'} \geq \Delta^1_{i,t,t'+1} \qquad \forall i,\ t\text{=1,2,..,}T\text{-1 and } t'\text{=}t,t\text{+1,..,}T\text{-1}. \qquad (5.112)$$

$$X^1_{i,t,t'} \geq \mathcal{E}_d - \tfrac{C_{t'}}{a_i}(1 - \Delta^1_{i,t,t'}) \qquad \forall i,\ \forall t \text{ and } t'\text{=}t. \qquad (5.113)$$

$$a_i\, X^1_{i,t,t'} \leq C_{t'}\, \Delta^1_{i,t,t'} \qquad \forall i, \forall t \text{ and } t'\text{=}t,t\text{+1,..,}T. \qquad (5.114)$$

$$s^1_{i,t} = ST_{0,i}\, \delta^1_{0,i,t} \qquad \forall i \text{ and } t\text{=1}. \qquad (5.115)$$

$$s^1_{i,t} = \sum_{i'=1}^{N} ST_{i',i}\, \delta^1_{i',i,t} \qquad \forall i \text{ and } t\text{=2,3,..,}T. \qquad (5.116)$$

$$s^1_{i,t} \leq (C_t - \mathcal{E})\delta^1_{0,i,t} \qquad \forall i \text{ and } t\text{=1}. \qquad (5.117)$$

$$s^1_{i,t} \leq (C_t - \mathcal{E}) \sum_{i'=1}^{N} \delta^1_{i',i,t} \qquad \forall i \text{ and } t\text{=2,3,..,}T. \qquad (5.118)$$

/* Constraints (5.119)–(5.124) correspond to a possible setup occurring anywhere in period t (but not at the beginning or the end of period t), with the production starting in period t. The setup time of the product being set up (using these constraints) should be less than the capacity of period t */

$$\Delta^2_{i,t,t} = \sum_{i'=1}^{N} \delta^2_{i',i,t} \qquad \forall i \text{ and } \forall t. \qquad (5.119)$$

$$\Delta^2_{i,t,t'} \geq \Delta^2_{i,t,t'+1} \qquad \forall i,\ t\text{=1,2,..,}T\text{-1 and } t'\text{=}t,t\text{+1,..,}T\text{-1}. \qquad (5.120)$$

$$X^2_{i,t,t'} \geq \mathcal{E}_d - \tfrac{C_{t'}}{a_i}(1 - \Delta^2_{i,t,t'}) \qquad \forall i,\ \forall t \text{ and } t'\text{=}t. \qquad (5.121)$$

$$a_i\, X^2_{i,t,t'} \leq C_{t'}\, \Delta^2_{i,t,t'} \qquad \forall i,\ \forall t \text{ and } t'\text{=}t,t\text{+1,..,}T. \qquad (5.122)$$

$$s^2_{i,t} = \sum_{i'=1}^{N} ST_{i',i}\, \delta^2_{i',i,t} \qquad \forall i \text{ and } \forall t. \qquad (5.123)$$

$$s^2_{i,t} \leq (C_t - \mathcal{E}) \sum_{i'=1}^{N} \delta^2_{i',i,t} \qquad \forall i \text{ and } \forall t. \qquad (5.124)$$

/* Constraints (5.125)–(5.130) correspond to a possible end-of-period setup occurring in period t, with the production starting in period $t+1$. The setup time of the product being set up (using these constraints) should be less than the capacity of that period */

$$\Delta^3_{i,t,t+1} = \sum_{i'=1}^{N} \delta^3_{i',i,t} \qquad \forall i \text{ and } t\text{=}1,2,..,T\text{-}1. \tag{5.125}$$

$$\Delta^3_{i,t,t'} \geq \Delta^3_{i,t,t'+1} \qquad \forall i,\ t\text{=}1,2,..,T\text{-}2 \text{ and } t'\text{=}t\text{+}1,t\text{+}2,..,T\text{-}1. \tag{5.126}$$

$$X^3_{i,t,t'} \geq \mathcal{E}_d - \frac{C_{t'}}{a_i}(1-\Delta^3_{i,t,t'}) \qquad \forall i,\ t\text{=}1,2,..,T\text{-}1 \text{ and } t'\text{=}t\text{+}1. \tag{5.127}$$

$$a_i\, X^3_{i,t,t'} \leq C_{t'}\, \Delta^3_{i,t,t'} \qquad \forall i,\ t\text{=}1,2,..,T\text{-}1 \text{ and } t'\text{=}t\text{+}1,t\text{+}2,..,T. \tag{5.128}$$

$$s^3_{i,t} = \sum_{i'=1}^{N} ST_{i',i}\, \delta^3_{i',i,t} \qquad \forall i \text{ and } t\text{=}1,2,..,T\text{-}1. \tag{5.129}$$

$$s^3_{i,t} \leq (C_t - \mathcal{E}) \sum_{i'=1}^{N} \delta^3_{i',i,t} \qquad \forall i \text{ and } t\text{=}1,2,..,T\text{-}1. \tag{5.130}$$

/* Constraints (5.131)–(5.139) correspond to a possible complete setup done in period t (with the setup started exactly in the beginning of period t, and completed exactly at the end of period t), with the production starting in period $t+1$. The setup time of the product being set up (using these constraints) should be equal to the capacity of that period */

$$\Delta^4_{i,t,t+1} = \delta^4_{0,i,t} \qquad \forall i \text{ and } t\text{=}1. \tag{5.131}$$

$$\Delta^4_{i,t,t+1} = \sum_{i'=1}^{N} \delta^4_{i',i,t} \qquad \forall i \text{ and } t\text{=}2,3,..,T\text{-}1. \tag{5.132}$$

$$\Delta^4_{i,t,t'} \geq \Delta^4_{i,t,t'+1} \qquad \forall i,\ t\text{=}1,2,..,T\text{-}2 \text{ and } t'\text{=}t\text{+}1,t\text{+}2,..,T\text{-}1. \tag{5.133}$$

$$X^4_{i,t,t'} \geq \mathcal{E}_d - \frac{C_{t'}}{a_i}(1-\Delta^4_{i,t,t'}) \qquad \forall i,\ t\text{=}1,2,..,T\text{-}1 \text{ and } t'\text{=}t\text{+}1. \tag{5.134}$$

$$a_i\, X^4_{i,t,t'} \leq C_{t'}\, \Delta^4_{i,t,t'} \qquad \forall i,\ t\text{=}1,2,..,T\text{-}1 \text{ and } t'\text{=}t\text{+}1,t\text{+}2,..,T. \tag{5.135}$$

$$s^4_{i,t} = ST_{0,i}\, \delta^4_{0,i,t} \qquad \forall i \text{ and } t\text{=}1. \tag{5.136}$$

$$s^4_{i,t} = \sum_{i'=1}^{N} ST_{i',i}\, \delta^4_{i',i,t} \qquad \forall i \text{ and } t\text{=}2,3,..,T\text{-}1. \tag{5.137}$$

$$s^4_{i,t} = C_t \delta^4_{0,i,t} \qquad \forall i \text{ and } t\text{=}1. \tag{5.138}$$

$$s^4_{i,t} = C_t \sum_{i'=1}^{N} \delta^4_{i',i,t} \qquad \forall i \text{ and } t\text{=}2,3,..,T\text{-}1. \tag{5.139}$$

/* Constraints (5.140)–(5.156) correspond to a possible setup crossover across a number of periods with the production starting in the same period where the setup ends. These constraints hold for the product which has its setup starting in the beginning of period t and ending in the middle of some future period t'. The setup time of the product being set up (using these constraints) should be greater than the capacity of the period in which the setup is initiated */

$$\Omega^5_{i,t,t',t} = \delta^5_{0,i,t,t'} \quad \forall i, t\text{=}1 \text{ and } t'\text{=}t\text{+}1,t\text{+}2,..,T. \tag{5.140}$$

$$\Omega^5_{i,t,t',t} = \sum_{i'=1}^{N} \delta^5_{i',i,t,t'} \quad \forall i, t\text{=}2,3,..,T\text{-}1 \text{ and } t'\text{=}t\text{+}1,t\text{+}2,..,T. \tag{5.141}$$

$$\Omega^5_{i,t,t',t''} = \Omega^5_{i,t,t',t''+1} \quad \forall i, t\text{=}1,2,..,T\text{-}1, t'\text{=}t\text{+}1,t\text{+}2,..,T \text{ and } t''\text{=}t, t\text{+}1,..,t'\text{-}1. \tag{5.142}$$

$$\Omega^5_{i,t,t',t''} = \Delta^5_{i,t,t',t''} \quad \forall i, t\text{=}1,2,..,T\text{-}1, t'\text{=}t\text{+}1,t\text{+}2,..,T \text{ and } t''\text{=}t'. \tag{5.143}$$

$$\Delta^5_{i,t,t',t''} \geq \Delta^5_{i,t,t',t''+1} \quad \forall i, t\text{=}1,2,..,T\text{-}2, t'\text{=}t\text{+}1,t\text{+}2,..,T\text{-}1 \text{ and } t''\text{=}t',t'\text{+}1,..,T\text{-}1. \tag{5.144}$$

$$\sum_{t'=t+1}^{T} \Omega^5_{i,t,t',t} \leq 1 \quad \forall i \text{ and } t\text{=}1,2,..,T\text{-}1. \tag{5.145}$$

$$\sum_{t''=t}^{t'} \Omega^5_{i,t,t',t''} = (t' - t + 1)\delta^5_{0,i,t,t'} \quad \forall i, t\text{=}1 \text{ and } t'\text{=}t\text{+}1,t\text{+}2,..,T. \tag{5.146}$$

$$\sum_{t''=t}^{t'} \Omega^5_{i,t,t',t''} = (t' - t + 1) \sum_{i'=1}^{N} \delta^5_{i',i,t,t'} \quad \forall i, t\text{=}2,3,..,T\text{-}1 \text{ and } t'\text{=}t\text{+}1,t\text{+}2,..,T. \tag{5.147}$$

$$\Omega^5_{i,t,t',t''} = 0 \quad \forall i, t\text{=}1,2,..,T\text{-}2, t'\text{=}t\text{+}1,t\text{+}2,..,T\text{-}1 \text{ and } t''\text{=}t'\text{+}1,t'\text{+}2,..,T. \tag{5.148}$$

$$\Delta^5_{i,t,t',T+1} = 0 \quad \forall i, t\text{=}1,2,..,T\text{-}1 \text{ and } t'\text{=}t\text{+}1,t\text{+}2,..,T. \tag{5.149}$$

$$X^5_{i,t,t',t''} \geq \mathcal{E}_d - \frac{C_{t''}}{a_i}(1 - \Delta^5_{i,t,t',t''}) \quad \forall i, t\text{=}1,2,..,T\text{-}1, t'\text{=}t\text{+}1,t\text{+}2,..,T \text{ and } t''\text{=}t'. \tag{5.150}$$

$$a_i\, X^5_{i,t,t',t''} \leq C_{t''}\, \Delta^5_{i,t,t',t''} \quad \forall i, t\text{=}1,2,..,T\text{-}1, t'\text{=}t\text{+}1,t\text{+}2,..,T \text{ and } t''\text{=}t',t'\text{+}1,..,T. \tag{5.151}$$

$$s^5_{i,t,t',t''} \geq \mathcal{E} - C_{t''}(1 - \Omega^5_{i,t,t',t''}) \quad \forall i, t\text{=}1,2,..,T\text{-}1, t'\text{=}t\text{+}1,t\text{+}2,..,T \text{ and } t''\text{=}t,t\text{+}1,..,t'. \tag{5.152}$$

$$\sum_{t''=t}^{t'} s^5_{i,t,t',t''} = ST_{0,i}\, \delta^5_{0,i,t,t'} \quad \forall i, t\text{=}1 \text{ and } t'\text{=}t\text{+}1,t\text{+}2,..,T. \tag{5.153}$$

$$\sum_{t''=t}^{t'} s^5_{i,t,t',t''} = \sum_{i'=1}^{N} ST_{i',i}\, \delta^5_{i',i,t,t'} \quad \forall i, t\text{=}2,3,..,T\text{-}1 \text{ and } t'\text{=}t\text{+}1,t\text{+}2,..,T. \tag{5.154}$$

$$s^5_{i,t,t',t''} \le (C_{t''} - \mathcal{E})\,\Omega^5_{i,t,t',t''}$$ $\forall i$, t=1,2,..,T-1, t'=t+1,t+2,..,T and t''=t'. (5.155)

$$s^5_{i,t,t',t''} = C_{t''}\Omega^5_{i,t,t',t''}$$ $\forall i$, t=1,2,..,T-1, t'=t+1,t+2,..,T and t''=t,t+1,..,t'-1. (5.156)

/* Constraints (5.157)–(5.172) represent a possible setup crossover across a number of periods with the production starting in period $t'+1$ when the setup ends at the end of period t'. These constraints hold for the product which has its setup starting in the beginning of period t and ends exactly at the end of some future period t'. The setup time of the product being set up (using these constraints) should be greater than the capacity of the period in which the setup is initiated */

$$\Omega^6_{i,t,t',t} = \delta^6_{0,i,t,t'}$$ $\forall i$, t=1 and t'=t+1,t+2,..,T-1. (5.157)

$$\Omega^6_{i,t,t',t} = \sum_{i'=1}^{N} \delta^6_{i',i,t,t'}$$ $\forall i$, t=2,3,..,T-2 and t'=t+1,t+2,..,T-1. (5.158)

$$\Omega^6_{i,t,t',t''} = \Omega^6_{i,t,t',t''+1}$$ $\forall i$, t=1,2,..,T-2, t'=t+1,t+2,..,T-1 and t''=t,t+1,..,t'-1. (5.159)

$$\Omega^6_{i,t,t',t''} = \Delta^6_{i,t,t',t''+1}$$ $\forall i$, t=1,2,..,T-2, t'=t+1,t+2,..,T-1 and t''=t'. (5.160)

$$\Delta^6_{i,t,t',t''} \ge \Delta^6_{i,t,t',t''+1}$$ $\forall i$, t=1,2,..,T-3, t'=t+1,t+2,..,T-2 and t''=t'+1,t'+2,..,T-1. (5.161)

$$\sum_{t'=t+1}^{T-1} \Omega^6_{i,t,t',t} \le 1$$ $\forall i$ and t=1,2,..,T-2. (5.162)

$$\sum_{t''=t}^{t'} \Omega^6_{i,t,t',t''} = (t'-t+1)\delta^6_{0,i,t,t'}$$ $\forall i$, t=1 and t'=t+1,t+2,..,T-1. (5.163)

$$\sum_{t''=t}^{t'} \Omega^6_{i,t,t',t''} = (t'-t+1)\sum_{i'=1}^{N} \delta^6_{i',i,t,t'}$$ $\forall i$, t=2,3,..,T-2 and t'=t+1,t+2,..,T-1. (5.164)

$$\Omega^6_{i,t,t',t''} = 0$$ $\forall i$, t=1,2,..,T-2, t'=t+1,t+2,..,T-1 and t''=t'+1,t'+2,..,T. (5.165)

$$\Delta^6_{i,t,t',T+1} = 0$$ $\forall i$, t=1,2,..,T-2 and t'=t+1,t+2,..,T-1. (5.166)

$$X^6_{i,t,t',t''} \ge \mathcal{E}_d - \frac{C_{t''}}{a_i}(1-\Delta^6_{i,t,t',t''})$$ $\forall i$, t=1,2,..,T-2, t'=t+1,t+2,..,T-1 and t''=t'+1. (5.167)

$$a_i X^6_{i,t,t',t''} \leq C_{t''} \Delta^6_{i,t,t',t''}$$ $\forall i$, t=1,2,..,T-2, t'=t+1,t+2,..,T-1 and t''=t'+1,t'+2,..,T. (5.168)

$$s^6_{i,t,t',t''} \geq \mathcal{E} - C_{t''}(1 - \Omega^6_{i,t,t',t''})$$ $\forall i$, t=1,2,..,T-2, t'=t+1,t+2,..,T-1 and t''=t,t+1,..,t'. (5.169)

$$\sum_{t''=t}^{t'} s^6_{i,t,t',t''} = ST_{0,i}\, \delta^6_{0,i,t,t'}$$ $\forall i$, t=1 and t'=t+1,t+2,..,T-1. (5.170)

$$\sum_{t''=t}^{t'} s^6_{i,t,t',t''} = \sum_{i'=1}^{N} ST_{i',i}\, \delta^6_{i',i,t,t'}$$ $\forall i$, t=2,3,..,T-2 and t'=t+1,t+2,..,T-1. (5.171)

$$s^6_{i,t,t',t''} = C_{t''} \Omega^6_{i,t,t',t''}$$ $\forall i$, t=1,2,..,T-2, t'=t+1,t+2,..,T-1 and t''=t,t+1,..,t'. (5.172)

/* Constraints (5.173)–(5.186) represent a possible setup crossover across a number of periods with the production starting in the same period where the setup ends. These constraints hold for the product which has its setup starting in the middle of period t and ends in the middle of some future period t'. The setup time of the product being set up (using these constraints) can be greater/less than the capacity of the period in which the setup is initiated */

$$\Omega^7_{i,t,t',t} = \sum_{i'=1}^{N} \delta^7_{i',i,t,t'}$$ $\forall i$, t=1,2,..,T-1 and t'=t+1,t+2,..,T. (5.173)

$$\Omega^7_{i,t,t',t''} = \Omega^7_{i,t,t',t''+1}$$ $\forall i$, t=1,2,..,T-1, t'=t+1,t+2,..,T and t''=t,t+1,..,t'-1. (5.174)

$$\Omega^7_{i,t,t',t''} = \Delta^7_{i,t,t',t''}$$ $\forall i$, t=1,2,..,T-1, t'=t+1,t+2,..,T and t''=t'. (5.175)

$$\Delta^7_{i,t,t',t''} \geq \Delta^7_{i,t,t',t''+1}$$ $\forall i$, t=1,2,..,T-2, t'=t+1,t+2,..,T-1 and t''=t',t'+1,..,T-1. (5.176)

$$\sum_{t'=t+1}^{T} \Omega^7_{i,t,t',t} \leq 1$$ $\forall i$ and t=1,2,..,T-1. (5.177)

$$\sum_{t''=t}^{t'} \Omega^7_{i,t,t',t''} = (t' - t + 1) \sum_{i'=1}^{N} \delta^7_{i',i,t,t'}$$ $\forall i$, t=1,2,..,T-1 and t'=t+1,t+2,..,T. (5.178)

$$\Omega^7_{i,t,t',t''} = 0$$ $\forall i$, t=1,2,..,T-2, t'=t+1,t+2,..,T-1 and t''=t'+1,t'+2,..,T. (5.179)

$$\Delta^7_{i,t,t',T+1} = 0$$ $\forall i$, t=1,2,..,T-1 and t'=t+1,t+2,..,T. (5.180)

$$X^7_{i,t,t',t''} \geq \mathcal{E}_d - \frac{C_{t''}}{a_i}(1 - \Delta^7_{i,t,t',t''})$$ $\forall i$, t=1,2,..,T-1, t'=t+1,t+2,..,T and t''=t'. (5.181)

$$a_i\, X^7_{i,t,t',t''} \le C_{t''}\, \Delta^7_{i,t,t',t''}$$ $\forall i$, t=1,2,..,T-1, t'=t+1,t+2,..,T and t''=t',t'+1,..,T. (5.182)

$$s^7_{i,t,t',t''} \ge \mathcal{E} - C_{t''}(1 - \Omega^7_{i,t,t',t''})$$ $\forall i$, t=1,2,..,T-1, t'=t+1,t+2,..,T and t''=t,t+1,..,t'. (5.183)

$$\sum_{t''=t}^{t'} s^7_{i,t,t',t''} = \sum_{i'=1}^{N} ST_{i',i}\, \delta^7_{i',i,t,t'}$$ $\forall i$, t=1,2,..,T-1 and t'=t+1,t+2,..,T. (5.184)

$$s^7_{i,t,t',t''} \le (C_{t''} - \mathcal{E})\, \Omega^7_{i,t,t',t''}$$ $\forall i$, t=1,2,..,T-1, t'=t+1,t+2,..,T and t''=t & t'. (5.185)

$$s^7_{i,t,t',t''} = C_{t''}\Omega^7_{i,t,t',t''}$$ $\forall i$, t=1,2,..,T-1, t'=t+1,t+2,..,T and t''=t+1,t+2,..,t'-1. (5.186)

/* Constraints (5.187)–(5.200) represent a possible setup crossover across a number of periods with the production starting in period $t' + 1$ when the setup ends at the end of period t'. These constraints hold for the product which has its setup starting in the middle of period t and ends exactly at the end of some future period t'. The setup time of the product being set up (using these constraints) should be greater than the capacity of the period in which the setup is initiated */

$$\Omega^8_{i,t,t',t} = \sum_{i'=1}^{N} \delta^8_{i',i,t,t'}$$ $\forall i$, t=1,2,..,T-2 and t'=t+1,t+2,..,T-1. (5.187)

$$\Omega^8_{i,t,t',t''} = \Omega^8_{i,t,t',t''+1}$$ $\forall i$, t=1,2,..,T-2, t'=t+1,t+2,..,T-1 and t''=t,t+1,..,t'-1. (5.188)

$$\Omega^8_{i,t,t',t''} = \Delta^8_{i,t,t',t''+1}$$ $\forall i$, t=1,2,..,T-2, t'=t+1,t+2,..,T-1 and t''=t'. (5.189)

$$\Delta^8_{i,t,t',t''} \ge \Delta^8_{i,t,t',t''+1}$$ $\forall i$, t=1,2,..,T-3, t'=t+1,t+2,..,T-2 and t''=t'+1,t'+2,..,T-1. (5.190)

$$\sum_{t'=t+1}^{T-1} \Omega^8_{i,t,t',t} \le 1$$ $\forall i$ and t=1,2,..,T-2. (5.191)

$$\sum_{t''=t}^{t'} \Omega^8_{i,t,t',t''} = (t' - t + 1) \sum_{i'=1}^{N} \delta^8_{i',i,t,t'}$$ $\forall i$, t=1,2,..,T-2 and t'=t+1,t+2,..,T-1. (5.192)

$$\Omega^8_{i,t,t',t''} = 0$$ $\forall i$, t=1,2,..,T-2, t'=t+1,t+2,..,T-1 and t''=t'+1,t'+2,..,T. (5.193)

$$\Delta^8_{i,t,t',T+1} = 0$$ $\forall i$, t=1,2,..,T-2 and t'=t+1,t+2,..,T-1. (5.194)

$X^8_{i,t,t',t''} \geq \mathcal{E}_d - \frac{C_{t''}}{a_i}(1 - \Delta^8_{i,t,t',t''})$ $\forall i$, t=1,2,..,T-2, t'=t+1,t+2,..,T-1 and t''=t'+1. (5.195)

$a_i\, X^8_{i,t,t',t''} \leq C_{t''}\, \Delta^8_{i,t,t',t''}$ $\forall i$, t=1,2,..,T-2, t'=t+1,t+2,..,T-1 and t''=t'+1,t'+2,..,T. (5.196)

$s^8_{i,t,t',t''} \geq \mathcal{E} - C_{t''}(1 - \Omega^8_{i,t,t',t''})$ $\forall i$, t=1,2,..,T-2, t'=t+1,t+2,..,T-1 and t''=t,t+1,..,t'. (5.197)

$\sum_{t''=t}^{t'} s^8_{i,t,t',t''} = \sum_{i'=1}^{N} ST_{i',i}\, \delta^8_{i',i,t,t'}$ $\forall i$, t=1,2,..,T-2 and t'=t+1,t+2,..,T-1. (5.198)

$s^8_{i,t,t',t''} = C_{t''}\Omega^8_{i,t,t',t''}$ $\forall i$, t=1,2,..,T-2, t'=t+1,t+2,..,T-1 and t''=t+1,t+2,..,t'. (5.199)

$s^8_{i,t,t',t''} \leq (C_{t''} - \mathcal{E})\Omega^8_{i,t,t',t''}$ $\forall i$, t=1,2,..,T-2, t'=t+1,t+2,..,T-1 and t''=t. (5.200)

/* Constraint (5.201) restricts only one product to be setup in the beginning of period 1 */

$$\sum_{i=1}^{N}(\delta^1_{0,i,1} + \delta^4_{0,i,1} + \sum_{t'=2}^{T}\delta^5_{0,i,1,t'} + \sum_{t'=2}^{T-1}\delta^6_{0,i,1,t'}) = 1. \quad (5.201)$$

/* Constraint (5.202) indicates that only one product can be tracked with respect to the beginning of every period t */

$\sum_{i=1}^{N}\alpha_{i,t} = 1,$ t=1,2,..,T+1. (5.202)

/* Constraint (5.203) indicates that at most one product can be tracked with respect to the end of every period t */

$\sum_{i=1}^{N}\beta_{i,t} \leq 1,$ t=1,2,..,T-1. (5.203)

/* Constraints (5.204)–(5.208) indicate that at most one product can be tracked as the product last setup with respect to the end of every period t */

$\beta_{i,t} - \alpha_{i,t} \leq \Delta^2_{i,t,t}$ $\forall i$ and t=1,2,..,T-1. (5.204)

$$\beta_{i,t} + \Delta^2_{i',t,t} \leq 1 \qquad \forall i, \forall i', i' \neq i \text{ and } t=1,2,..,T\text{-}1. \qquad (5.205)$$

$$\beta_{i,1} + \sum_{i'=1}^{N}(\sum_{i''=1}^{N}(\delta^1_{i'',i',2} + \delta^4_{i'',i',2}) + \Delta^1_{i',1,2}$$
$$+\Delta^2_{i',1,2} + \Delta^3_{i',1,2} + \Delta^4_{i',1,2} + \sum_{t'=2}^{T}\Omega^5_{i',1,t',2}$$
$$+\sum_{t'=2}^{T-1}\Omega^6_{i',1,t',2} + \sum_{t'=2}^{T}\Omega^7_{i',1,t',2}$$
$$+\sum_{t'=2}^{T-1}\Omega^8_{i',1,t',2} + \sum_{t'=3}^{T}\Omega^5_{i',2,t',2}$$
$$+\sum_{t'=3}^{T-1}\Omega^6_{i',2,t',2}) \leq 1 \qquad \forall i. \qquad (5.206)$$

$$\beta_{i,t-1} + \sum_{i'=1}^{N}(\sum_{i''=1}^{N}(\delta^1_{i'',i',t} + \delta^4_{i'',i',t})$$
$$+\sum_{t'=1}^{t-1}(\Delta^1_{i',t',t} + \Delta^2_{i',t',t} + \Delta^3_{i',t',t} + \Delta^4_{i',t',t})$$
$$+\sum_{t''=t'+1}^{t-1}\sum_{t'=1}^{t-2}(\Delta^5_{i',t',t'',t} + \Delta^6_{i',t',t'',t}$$
$$+\Delta^7_{i',t',t'',t} + \Delta^8_{i',t',t'',t}) + \sum_{t''=t}^{T}\Omega^5_{i',1,t'',t}$$
$$+\sum_{t''=t}^{T-1}\Omega^6_{i',1,t'',t} + \sum_{t''=t}^{T}\sum_{t'=2}^{t-1}\Omega^5_{i',t',t'',t}$$
$$+\sum_{t''=t}^{T-1}\sum_{t'=2}^{t-1}\Omega^6_{i',t',t'',t} + \sum_{t''=t+1}^{T}\Omega^5_{i',t,t'',t}$$
$$+\sum_{t''=t+1}^{T-1}\Omega^6_{i',t,t'',t} + \sum_{t''=t}^{T}\sum_{t'=1}^{t-1}\Omega^7_{i',t',t'',t}$$
$$+\sum_{t''=t}^{T-1}\sum_{t'=1}^{t-1}\Omega^8_{i',t',t'',t}) \leq 1 \qquad \forall i \text{ and } t=3,4,..,T\text{-}1. \qquad (5.207)$$

$$\beta_{i,T-1} + \sum_{i'=1}^{N}(\sum_{i''=1}^{N}\delta^1_{i'',i',T} + \sum_{t'=1}^{T-1}(\Delta^1_{i',t',T}$$
$$+\Delta^2_{i',t',T} + \Delta^3_{i',t',T} + \Delta^4_{i',t',T})$$
$$+\sum_{t''=t'+1}^{T-1}\sum_{t'=1}^{T-2}(\Delta^5_{i',t',t'',T} + \Delta^6_{i',t',t'',T}+$$
$$\Delta^7_{i',t',t'',T} + \Delta^8_{i',t',t'',T})$$
$$+\sum_{t'=1}^{T-1}\Omega^5_{i',t',T,T} + \sum_{t'=1}^{T-1}\Omega^7_{i',t',T,T}) \leq 1 \qquad \forall i. \qquad (5.208)$$

/* Constraint (5.209)–(5.212) tracks the product last setup with respect to the beginning of every period t */

$$\alpha_{i,1} \leq \delta^1_{0,i,1} + \delta^4_{0,i,1}$$
$$+\sum_{t'=2}^{T}\Omega^5_{i,1,t',1} + \sum_{t'=2}^{T-1}\Omega^6_{i,1,t',1} \qquad \forall i. \qquad (5.209)$$

$$\alpha_{i,2}-\beta_{i,1} \leq \sum_{i'=1}^{N}(\delta^1_{i',i,2}+\delta^4_{i',i,2})+\Delta^1_{i,1,2}+\Delta^2_{i,1,2}$$

$$+\Delta^3_{i,1,2}+\Delta^4_{i,1,2}+\sum_{t'=2}^{T}\Omega^5_{i,1,t',2}$$

$$+\sum_{t'=2}^{T-1}\Omega^6_{i,1,t',2}+\sum_{t'=2}^{T}\Omega^7_{i,1,t',2}$$

$$+\sum_{t'=2}^{T-1}\Omega^8_{i,1,t',2}+\sum_{t'=3}^{T}\Omega^5_{i,2,t',2}+\sum_{t'=3}^{T-1}\Omega^6_{i,2,t',2} \quad \forall i. \tag{5.210}$$

$$\alpha_{i,t}-\beta_{i,t-1} \leq \sum_{i'=1}^{N}(\delta^1_{i',i,t}+\delta^4_{i',i,t})+$$

$$\sum_{t'=1}^{t-1}(\Delta^1_{i,t',t}+\Delta^2_{i,t',t}+\Delta^3_{i,t',t}+\Delta^4_{i,t',t})$$

$$+\sum_{t''=t'+1}^{t-1}\sum_{t'=1}^{t-2}(\Delta^5_{i,t',t'',t}$$

$$\Delta^6_{i,t',t'',t}+\Delta^7_{i,t',t'',t}+\Delta^8_{i,t',t'',t})$$

$$+\sum_{t''=t}^{T}\Omega^5_{i,1,t'',t}+\sum_{t''=t}^{T-1}\Omega^6_{i,1,t'',t}$$

$$+\sum_{t''=t}^{T}\sum_{t'=2}^{t-1}\Omega^5_{i,t',t'',t}+\sum_{t'=t}^{T-1}\sum_{t''=2}^{t-1}\Omega^6_{i,t',t'',t}$$

$$+\sum_{t''=t+1}^{T}\Omega^5_{i,t,t'',t}+\sum_{t''=t+1}^{T-1}\Omega^6_{i,t,t'',t}$$

$$+\sum_{t''=t}^{T}\sum_{t'=1}^{t-1}\Omega^7_{i,t',t'',t}+\sum_{t''=t}^{T-1}\sum_{t'=1}^{t-1}\Omega^8_{i,t',t'',t} \quad \forall i \text{ and } t=3,4,..,T\text{-}1. \tag{5.211}$$

$$\alpha_{i,T}-\beta_{i,T-1} \leq \sum_{i'=1}^{N}\delta^1_{i',i,T}+\sum_{t'=1}^{T-1}(\Delta^1_{i,t',T}$$

$$+\Delta^2_{i,t',T}+\Delta^3_{i,t',T}+\Delta^4_{i,t',T})$$

$$+\sum_{t''=t'+1}^{T-1}\sum_{t'=1}^{T-2}(\Delta^5_{i,t',t'',T}+\Delta^6_{i,t',t'',T}+$$

$$\Delta^7_{i,t',t'',T}+\Delta^8_{i,t',t'',T})$$

$$+\sum_{t'=1}^{T-1}\Omega^5_{i,t',T,T}+\sum_{t'=1}^{T-1}\Omega^7_{i,t',T,T} \quad \forall i. \tag{5.212}$$

/* Constraints (5.213)–(5.215) restrict a product i to be set up in period t with the help of only one type of setup indicator */

$$(\delta^1_{0,i,1}+\delta^4_{0,i,1})+\sum_{i'=1}^{N}(\delta^2_{i',i,1}+\delta^3_{i',i,1})$$

$$+\sum_{t'=2}^{T}\delta^5_{0,i,1,t'}+\sum_{t'=2}^{T-1}\delta^6_{0,i,1,t'}$$

$$+\sum_{i'=1}^{N}(\sum_{t'=2}^{T}\delta^7_{i',i,1,t'}+\sum_{t'=2}^{T-1}\delta^8_{i',i,1,t'}) \leq 1 \quad \forall i. \tag{5.213}$$

$\sum_{i'=1}^{N}(\delta^1_{i',i,t} + \delta^2_{i',i,t} + \delta^3_{i',i,t} + \delta^4_{i',i,t}$

$+\sum_{t'=t+1}^{T}\delta^5_{i',i,t,t'}$

$\sum_{t'=t+1}^{T-1}\delta^6_{i',i,t,t'} + \sum_{t'=t+1}^{T}\delta^7_{i',i,t,t'}$

$$+\sum_{t'=t+1}^{T-1}\delta^8_{i',i,t,t'}) \leq 1 \qquad \forall i \text{ and } t=2,3,..,T\text{-}1. \tag{5.214}$$

$$\sum_{i'=1}^{N}(\delta^1_{i',i,t} + \delta^2_{i',i,t}) \leq 1 \qquad \forall i \text{ and } t=T. \tag{5.215}$$

/* Constraints (5.216)–(5.218) indicate that only one type of setup can occur between periods t and $t+1$ for at most one product. When a product i is supposed to be set up, only one of the following types of setup can occur in the border of periods t and $t+1$, i.e., an end-of period setup in period $t-1$ using $\delta^3_{i',i,t-1}$; or a beginning of period setup in period t using $\delta^1_{i',i,t}$; or a complete setup in period $t-1$ using $\delta^4_{i',i,t-1}$; or a complete setup in period t using $\delta^4_{i',i,t}$; or a setup crossover between periods t and $t+1$ which ends in some period $t' \geq t+1$ using $\delta^5_{i',i,t,t'}$ or $\delta^6_{i',i,t,t'}$ or $\delta^7_{i',i,t,t'}$ or $\delta^8_{i',i,t,t'}$ */

$\sum_{i'=1}^{N}\sum_{i=1}^{N}(\delta^1_{i',i,2}+\delta^3_{i',i,1}+\delta^4_{i',i,2}+\sum_{t'=2}^{T}\delta^7_{i',i,1,t'}$

$+\sum_{t'=2}^{T-1}\delta^8_{i',i,1,t'}) + \sum_{i=1}^{N}(\delta^4_{0,i,1} + \sum_{t'=2}^{T}\delta^5_{0,i,1,t'}$

$$+\sum_{t'=2}^{T-1}\delta^6_{0,i,1,t'}) \leq 1. \tag{5.216}$$

$\sum_{i'=1}^{N}\sum_{i=1}^{N}(\delta^1_{i',i,t} + \delta^3_{i',i,t-1} + \delta^4_{i',i,t-1} + \delta^4_{i',i,t}$

$+\sum_{t'=t+1}^{T}\delta^5_{i',i,t,t'} + \sum_{t'=t+1}^{T-1}\delta^6_{i',i,t,t'}$

$+\sum_{t'=1}^{t-2}\delta^8_{i',i,t',t-1} + \sum_{t'=2}^{t-2}\delta^6_{i',i,t',t-1}$

$+\sum_{t''=t}^{T}\sum_{t'=1}^{t-1}\delta^7_{i',i,t',t''} + \sum_{t''=t}^{T-1}\sum_{t'=1}^{t-1}\delta^8_{i',i,t',t''})$

$$+\sum_{i=1}^{N}(\delta^5_{0,i,1,t} + \delta^6_{0,i,1,t} + \delta^6_{0,i,1,t-1}) \leq 1, \qquad t=3,4,..,T\text{-}1. \tag{5.217}$$

$\sum_{i=1}^{N}(\delta^5_{0,i,1,T} + \delta^6_{0,i,1,T-1}) + \sum_{i'=1}^{N}\sum_{i=1}^{N}(\delta^1_{i',i,T}$

$+\delta^3_{i',i,T-1} + \delta^4_{i',i,T-1} + \sum_{t'=2}^{T-1}\delta^5_{i',i,t',T}$

$+\sum_{t'=1}^{T-1}\delta^7_{i',i,t',T} + \sum_{t'=1}^{T-2}\delta^8_{i',i,t',T-1}$

$$+\sum_{t'=2}^{T-2}\delta^6_{i',i,t',T-1}) \leq 1. \tag{5.218}$$

/* Constraints (5.219)–(5.222) represent the feasibility of production and setup with respect to capacity availability */

$$\sum_{i=1}^{N}(s_{i,1}^{1}+s_{i,1}^{2}+s_{i,1}^{3}+s_{i,1}^{4}+\sum_{t''=2}^{T}s_{i,1,t'',1}^{5}$$
$$+\sum_{t''=2}^{T-1}s_{i,1,t'',1}^{6}+\sum_{t''=2}^{T}s_{i,1,t'',1}^{7}$$
$$+\sum_{t''=2}^{T-1}s_{i,1,t'',1}^{8}+a_iX_{i,1,1}^{1}+a_iX_{i,1,1}^{2})\le C_1. \quad (5.219)$$

$$\sum_{i=1}^{N}(s_{i,2}^{1}+s_{i,2}^{2}+s_{i,2}^{3}+s_{i,2}^{4}+\sum_{t''=2}^{T}s_{i,1,t'',2}^{5}$$
$$+\sum_{t''=3}^{T}s_{i,2,t'',2}^{5}+\sum_{t''=2}^{T-1}s_{i,1,t'',2}^{6}$$
$$+\sum_{t''=3}^{T-1}s_{i,2,t'',2}^{6}+\sum_{t''=2}^{T}s_{i,1,t'',2}^{7}$$
$$+\sum_{t''=3}^{T}s_{i,2,t'',2}^{7}+\sum_{t''=2}^{T-1}s_{i,1,t'',2}^{8}$$
$$+\sum_{t''=3}^{T-1}s_{i,2,t'',2}^{8}+\sum_{t''=1}^{2}a_iX_{i,t'',2}^{1}$$
$$+\sum_{t''=1}^{2}a_iX_{i,t'',2}^{2}+a_iX_{i,1,2}^{3}$$
$$+a_iX_{i,1,2}^{4}+a_iX_{i,1,2,2}^{5}+a_iX_{i,1,2,2}^{7})\le C_2. \quad (5.220)$$

$$s_{i,t}^{1}+s_{i,t}^{2}+s_{i,t}^{3}+s_{i,t}^{8}+\sum_{t''=t}^{T}\sum_{t'=1}^{t-1}s_{i,t',t'',t}^{5}$$
$$+\sum_{t''=t}^{T-1}\sum_{t'=1}^{t-1}s_{i,t',t'',t}^{6}+\sum_{t''=t}^{T}\sum_{t'=1}^{t-1}s_{i,t',t'',t}^{7}$$
$$+\sum_{t''=t}^{T-1}\sum_{t'=1}^{t-1}s_{i,t',t'',t}^{8}+\sum_{t''=t+1}^{T}s_{i,t,t'',t}^{5}$$
$$+\sum_{t''=t+1}^{T-1}s_{i,t,t'',t}^{6}+\sum_{t''=t+1}^{T}s_{i,t,t'',t}^{7}$$
$$+\sum_{t''=t+1}^{T-1}s_{i,t,t'',t}^{8}+\sum_{t''=1}^{t}(a_iX_{i,t'',t}^{1}+a_iX_{i,t'',t}^{2})$$
$$+\sum_{t''=1}^{t-1}(a_iX_{i,t'',t}^{3}+a_iX_{i,t'',t}^{4})$$
$$+\sum_{t'''=t''+1}^{t}\sum_{t''=1}^{t-1}a_iX_{i,t'',t''',t}^{5}$$
$$+\sum_{t'''=t''+1}^{t-1}\sum_{t''=1}^{t-2}a_iX_{i,t'',t''',t}^{6}$$
$$+\sum_{t'''=t''+1}^{t}\sum_{t''=1}^{t-1}a_iX_{i,t'',t''',t}^{7}$$
$$+\sum_{t'''=t''+1}^{t-1}\sum_{t''=1}^{t-2}a_iX_{i,t'',t''',t}^{8}\le C_t, \qquad t=3,4,..,T\text{-}1. \quad (5.221)$$

$$\sum_{i=1}^{N}(s_{i,T}^{1}+s_{i,T}^{2}+\sum_{t'=1}^{T-1}s_{i,t',T,T}^{5}$$

$$+\sum_{t'=1}^{T-1}s_{i,t',T,T}^{7}+\sum_{t''=1}^{T}(a_iX_{i,t'',T}^{1}$$

$$+a_iX_{i,t'',T}^{2})+\sum_{t''=1}^{T-1}(a_iX_{i,t'',T}^{3}$$

$$+a_iX_{i,t'',T}^{4})+\sum_{t'''=t''+1}^{T}\sum_{t''=1}^{T-1}(a_iX_{i,t'',t''',T}^{5}$$

$$+a_iX_{i,t'',t''',T}^{7})+\sum_{t'''=t''+1}^{T-1}\sum_{t''=1}^{T-2}(a_iX_{i,t'',t''',T}^{6}$$

$$+a_iX_{i,t'',t''',T}^{8}))\leq C_T. \tag{5.222}$$

/* Constraints (5.223)–(5.226) indicate the flow balance constraints */

$$\alpha_{i,1}+\sum_{i'=1}^{N}\delta_{i',i,1}^{2}\geq\Delta_{i,1,2}^{1}+\Delta_{i,1,2}^{2}+\Delta_{i,1,2}^{4}+$$

$$\sum_{t'=2}^{T}\Omega_{i,1,t',2}^{5}+\sum_{t'=2}^{T-1}\Omega_{i,1,t',2}^{6}$$

$$+\sum_{i'=1}^{N}(\delta_{i,i',1}^{2}+\delta_{i,i',1}^{3}+\delta_{i,i',2}^{1}$$

$$+\delta_{i,i',2}^{4}+\sum_{t'=2}^{T-1}\delta_{i,i',1,t'}^{8}+\sum_{t'=2}^{T}\delta_{i,i',1,t'}^{7}$$

$$+\sum_{t'=3}^{T}\delta_{i,i',2,t'}^{5}+\sum_{t'=3}^{T-1}\delta_{i,i',2,t'}^{6})\qquad\forall i. \tag{5.223}$$

$$\alpha_{i,t}+\sum_{i'=1}^{N}\delta_{i',i,t}^{2}\geq\sum_{i'=1}^{N}(\delta_{i,i',t}^{2}+\delta_{i,i',t+1}^{1}$$

$$+\delta_{i,i',t+1}^{4}+\delta_{i,i',t}^{3}+\sum_{t'=t+1}^{T}\delta_{i,i',t,t'}^{7}$$

$$+\sum_{t'=t+1}^{T-1}\delta_{i,i',t,t'}^{8}+\sum_{t'=t+2}^{T}\delta_{i,i',t+1,t'}^{5}$$

$$+\sum_{t'=t+2}^{T-1}\delta_{i,i',t+1,t'}^{6}+\sum_{t'=1}^{t}(\Delta_{i,t',t+1}^{1}$$

$$+\Delta_{i,t',t+1}^{4}+\Delta_{i,t',t+1}^{2})+\sum_{t'=1}^{t-1}\Delta_{i,t',t+1}^{3}$$

$$+\sum_{t''=t'+1}^{t}\sum_{t'=1}^{t-1}(\Delta_{i,t',t'',t+1}^{5}+\Delta_{i,t',t'',t+1}^{6}$$

$$+\Delta_{i,t',t'',t+1}^{7}+\Delta_{i,t',t'',t+1}^{8})$$

$$+\sum_{t'''=t+1}^{T}\sum_{t''=1}^{t}\Omega_{i,t'',t''',t+1}^{5}$$

$$+\sum_{t'''=t+1}^{T-1}\sum_{t''=1}^{t}\Omega_{i,t'',t''',t+1}^{6}$$

$$+\sum_{t'''=t+1}^{T}\sum_{t''=1}^{t-1}\Omega_{i,t'',t''',t+1}^{7}$$

$$+\sum_{t'''=t+1}^{T-1}\sum_{t''=1}^{t-1}\Omega_{i,t'',t''',t+1}^{8})\qquad\forall i\text{ and }t=2,3,..,T\text{-}2. \tag{5.224}$$

$$\alpha_{i,T-1} + \sum_{i'=1}^{N} \delta^2_{i',i,T-1} \geq \sum_{i'=1}^{N} (\delta^2_{i,i',T-1}$$

$$+\delta^3_{i,i',T-1} + \delta^1_{i,i',T} + \delta^7_{i,i',t,T})$$

$$+\sum_{t'=1}^{T-1} (\Delta^1_{i,t',T} + \Delta^2_{i,t',T} + \Delta^4_{i,t',T})$$

$$+\sum_{t'=1}^{T-2} \Delta^3_{i,t',T} + \sum_{t''=t'+1}^{T-1} \sum_{t'=1}^{T-2} (\Delta^5_{i,t',t'',T}$$

$$+\Delta^6_{i,t',t'',T} + \Delta^7_{i,t',t'',T}$$

$$+\Delta^8_{i,t',t'',T}) + \sum_{t''=1}^{T-1} \Omega^5_{i,t'',T,T}$$

$$+\sum_{t''=1}^{T-2} \Omega^7_{i,t'',T,T} \qquad \forall i. \qquad (5.225)$$

$$\alpha_{i,T} + \sum_{i'=1}^{N} \delta^2_{i',i,T} = \sum_{i'=1}^{N} \delta^2_{i,i',T} + \alpha_{i,T+1} \qquad \forall i. \qquad (5.226)$$

/* Constraint (5.227) is the sub-tour elimination constraint */

$$U_{i,t} - U_{i',t} + N \times \delta^2_{i,i',t} \leq N - 1 \qquad i=0,1,..,N,\ i'=1,2,..,N,\ i \neq i' \text{ and } \forall t. \qquad (5.227)$$

/* Constraints (5.228)–(5.229) track the product last setup with respect to beginning of period t */

$$\delta^2_{0,i,t} \leq \alpha_{i,t} \qquad \forall i \text{ and } \forall t. \qquad (5.228)$$

$$\delta^2_{i,0,t} \leq \alpha_{i,t+1} \qquad \forall i \text{ and } \forall t. \qquad (5.229)$$

/* Constraints (5.230)–(5.233) indicate that a production carryover of any product i can take place in period t'' from period $t''-2$, only if the sum of the production times and setup times of all products in period $t''-1$ is equal to the capacity of $t''-1$ */

$$\sum_{i=1}^{N} (a_i\, X^1_{i,t,t''-1} + a_i\, X^2_{i,t,t''-1} + a_i\, X^3_{i,t,t''-1}$$

$$+a_i\, X^4_{i,t,t''-1} + \sum_{t'=t+1}^{t''-1} (a_i X^5_{i,t,t',t''-1}$$

$$+a_i X^7_{i,t,t',t''-1}) + \sum_{t'=t''-1}^{t''} (s^5_{i,t,t',t''-1}$$

$$+s^7_{i,t,t',t''-1}) + \sum_{t'=t''-1}^{t''-1} (s^6_{i,t,t',t''-1}$$

$$+s^{8}_{i,t,t',t''-1})) \leq C_{t''-1} + ((C_{t''-1} + 1)$$
$$\times(1 - \sum_{i=1}^{N}(\Delta^{1}_{i,t,t''} + \Delta^{2}_{i,t,t''} + \Delta^{3}_{i,t,t''} + \Delta^{4}_{i,t,t''}$$
$$+ \sum_{t'=t+1}^{t''}(\Delta^{5}_{i,t,t',t''} + \Delta^{7}_{i,t,t',t''})$$
$$+ \sum_{t'=t+1}^{t''-1}(\Delta^{6}_{i,t,t',t''} + \Delta^{8}_{i,t,t',t''})))), \quad t=1,2,..,T\text{-}2 \text{ and } t''=t+2. \tag{5.230}$$

$$\sum_{i=1}^{N}(a_i\, X^{1}_{i,t,t''-1} + a_i\, X^{2}_{i,t,t''-1} + a_i\, X^{3}_{i,t,t''-1}$$
$$+a_i\, X^{4}_{i,t,t''-1} + \sum_{t'=t+1}^{t''-1}(a_i X^{5}_{i,t,t',t''-1}$$
$$+a_i X^{7}_{i,t,t',t''-1}) + \sum_{t'=t''-1}^{t''}(s^{5}_{i,t,t',t''-1}$$
$$+s^{7}_{i,t,t',t''-1}) + \sum_{t'=t''-1}^{t''-1}(s^{6}_{i,t,t',t''-1}$$
$$+s^{8}_{i,t,t',t''-1})) \geq C_{t''-1} - ((C_{t''-1} + 1)$$
$$\times(1 - \sum_{i=1}^{N}(\Delta^{1}_{i,t,t''} + \Delta^{2}_{i,t,t''} + \Delta^{3}_{i,t,t''} + \Delta^{4}_{i,t,t''}$$
$$+ \sum_{t'=t+1}^{t''}(\Delta^{5}_{i,t,t',t''} + \Delta^{7}_{i,t,t',t''})$$
$$+ \sum_{t'=t+1}^{t''-1}(\Delta^{6}_{i,t,t',t''} + \Delta^{8}_{i,t,t',t''})))), \quad t=1,2,..,T\text{-}2 \text{ and } t''=t+2. \tag{5.231}$$

$$\sum_{i=1}^{N}(a_i\, X^{1}_{i,t,t''-1} + a_i\, X^{2}_{i,t,t''-1} + a_i\, X^{3}_{i,t,t''-1}$$
$$+a_i\, X^{4}_{i,t,t''-1} + \sum_{t'=t+1}^{t''-1}(a_i X^{5}_{i,t,t',t''-1}$$
$$+a_i X^{7}_{i,t,t',t''-1}) + \sum_{t'=t+1}^{t''-2}(a_i X^{6}_{i,t,t',t''-1}$$
$$+a_i X^{8}_{i,t,t',t''-1}) + \sum_{t'=t''-1}^{t''}(s^{5}_{i,t,t',t''-1}$$
$$+s^{7}_{i,t,t',t''-1}) + \sum_{t'=t''-1}^{t''-1}(s^{6}_{i,t,t',t''-1}$$
$$+s^{8}_{i,t,t',t''-1})) \leq C_{t''-1} + ((C_{t''-1} + 1)$$
$$\times(1 - \sum_{i=1}^{N}(\Delta^{1}_{i,t,t''} + \Delta^{2}_{i,t,t''} + \Delta^{3}_{i,t,t''} + \Delta^{4}_{i,t,t''}$$
$$+ \sum_{t'=t+1}^{t''}(\Delta^{5}_{i,t,t',t''} + \Delta^{7}_{i,t,t',t''})$$
$$+ \sum_{t'=t+1}^{t''-1}(\Delta^{6}_{i,t,t',t''} + \Delta^{8}_{i,t,t',t''})))), \quad t=1,2,..,T\text{-}3 \text{ and } t''=t+3,\ t+4,..,\ T. \tag{5.232}$$

$$\sum_{i=1}^{N}(a_i\,X_{i,t,t''-1}^{1}+a_i\,X_{i,t,t''-1}^{2}+a_i\,X_{i,t,t''-1}^{3}$$

$$+a_i\,X_{i,t,t''-1}^{4}+\sum_{t'=t+1}^{t''-1}(a_iX_{i,t,t',t''-1}^{5}$$

$$+a_iX_{i,t,t',t''-1}^{7})+\sum_{t'=t+1}^{t''-2}(a_iX_{i,t,t',t''-1}^{6}$$

$$+a_iX_{i,t,t',t''-1}^{8})+\sum_{t'=t''-1}^{t''}(s_{i,t,t',t''-1}^{5}$$

$$+s_{i,t,t',t''-1}^{7})+\sum_{t'=t''-1}^{t''-1}(s_{i,t,t',t''-1}^{6}$$

$$+s_{i,t,t',t''-1}^{8}))\geq C_{t''-1}-((C_{t''-1}+1)\times(1-$$

$$\sum_{i=1}^{N}(\Delta_{i,t,t''}^{1}+\Delta_{i,t,t''}^{2}+\Delta_{i,t,t''}^{3}+\Delta_{i,t,t''}^{4}$$

$$+\sum_{t'=t+1}^{t''}(\Delta_{i,t,t',t''}^{5}+\Delta_{i,t,t',t''}^{7})$$

$$+\sum_{t'=t+1}^{t''-1}(\Delta_{i,t,t',t''}^{6}+\Delta_{i,t,t',t''}^{8})))), \qquad t=1,2,..,T\text{-}2 \text{ and } t''=t+3,\ t+4,..,\ T. \qquad (5.233)$$

/* Constraints (5.234)–(5.235) represent the conditions for setting up a product only once in period t */

$$\sum_{i'=1}^{N}(\delta_{i',i,t}^{1}+\delta_{i',i,t}^{2}+\delta_{i',i,t}^{3}+\delta_{i',i,t}^{4}$$

$$+\sum_{t'=t+1}^{T}(\delta_{i',i,t,t'}^{5}+\delta_{i',i,t,t'}^{7})$$

$$+\sum_{t'=t+1}^{T-1}(\delta_{i',i,t,t'}^{6}+\delta_{i',i,t,t'}^{8}))\leq 1, \qquad \forall i \text{ and } t=1,2,..,T\text{-}2. \qquad (5.234)$$

$$\sum_{i'=1}^{N}(\delta_{i',i,T-1}^{1}+\delta_{i',i,T-1}^{2}+\delta_{i',i,T-1}^{3}+\delta_{i',i,T-1}^{4}$$

$$+\delta_{i',i,T-1,T}^{5}+\delta_{i',i,T-1,T}^{7})\leq 1 \qquad \forall i. \qquad (5.235)$$

/* Constraints (5.236)–(5.237) indicate that a production carryover can occur in period t from the preceding period for utmost one product, using utmost one type of production carryover indicator */

$$\sum_{i=1}^{N}(\Delta_{i,1,2}^{1}+\Delta_{i,1,2}^{2}+\Delta_{i,1,2}^{3}+\Delta_{i,1,2}^{4})\leq 1. \qquad (5.236)$$

$$\sum_{i=1}^{N}\sum_{t''=1}^{t-1}(\Delta^1_{i,t'',t}+\Delta^2_{i,t'',t}+\Delta^3_{i,t'',t}+\Delta^4_{i,t'',t})$$

$$+\sum_{i=1}^{N}\sum_{t'''=t''+1}^{t}\sum_{t''=1}^{t-1}(\Delta^5_{i,t'',t''',t}+\Delta^7_{i,t'',t''',t})$$

$$+\sum_{i=1}^{N}\sum_{t'''=t''+1}^{t-1}\sum_{t''=1}^{t-2}(\Delta^6_{i,t'',t''',t}$$

$$+\Delta^7_{i,t'',t''',t})\leq 1, \qquad t=3,4,..,T. \tag{5.237}$$

/* Constraints (5.238)–(5.240) show that if the setup of at least 1 product is initiated in period t (indicated by $delta_t$=1), then it means that no carryover of either setup or production of any product, with its setup having been initiated in any period up to $t-1$, can be carried over to period $t+1$ and later. This is so because if any product's setup is initiated in period t, then either the production or setup carryover with the setup initiated up to period $t-1$ and carried over to period $t+1$ and later is not feasible */

$$\sum_{i=1}^{N}\sum_{t''=1}^{t-1}\sum_{t'''=t+1}^{T}(\Omega^5_{i,t'',t''',t'''}+\Omega^6_{i,t'',t''',t'''}$$

$$+\Omega^7_{i,t'',t''',t'''}+\Omega^8_{i,t'',t''',t'''}+\Delta^1_{i,t'',t'''}+\Delta^2_{i,t'',t'''}$$

$$+\Delta^3_{i,t'',t'''}+\Delta^4_{i,t'',t'''}+\sum_{t''''=t''+1}^{t}(\Delta^5_{i,t'',t'''',t'''}$$

$$\Delta^6_{i,t'',t'''',t'''}+\Delta^7_{i,t'',t'''',t'''}+\Delta^8_{i,t'',t'''',t'''}))$$

$$\leq 6\times N\times T^2(1-delta_t), \qquad t=2,3,..,T\text{-}1. \tag{5.238}$$

$$\sum_{i=1}^{N}\sum_{\substack{i'=1\\ i\neq i'}}^{N}(\delta^1_{i',i,t}+\delta^2_{i',i,t}+\delta^3_{i',i,t}+\delta^4_{i',i,t}$$

$$+\sum_{t'=t+1}^{T}(\delta^5_{i',i,t,t'}+\delta^7_{i',i,t,t'})+\sum_{t'=t+1}^{T-1}(\delta^6_{i',i,t,t'}$$

$$+\delta^8_{i',i,t,t'}))\leq N\times delta_t, \qquad t=2,3,..,T\text{-}1. \tag{5.239}$$

$$delta_t\leq\sum_{i=1}^{N}\sum_{\substack{i'=1\\ i\neq i'}}^{N}(\delta^1_{i',i,t}+\delta^2_{i',i,t}+\delta^3_{i',i,t}+\delta^4_{i',i,t}$$

$$+\sum_{t'=t+1}^{T}(\delta^5_{i',i,t,t'}+\delta^7_{i',i,t,t'})+\sum_{t'=t+1}^{T-1}(\delta^6_{i',i,t,t'}$$

$$+\delta^8_{i',i,t,t'})), \qquad t=2,3,..,T\text{-}1. \tag{5.240}$$

/* Constraints (5.241)–(5.252) represent the inventory balance constraints */

$$I_{i,1}-B_{i,1}=X^1_{i,1,1}+X^2_{i,1,1}-d_{i,1} \qquad \forall i. \tag{5.241}$$

$$I_{i,2} - B_{i,2} = X^1_{i,1,1} + X^1_{i,1,2} + X^1_{i,2,2} + X^2_{i,1,1}$$
$$+X^2_{i,1,2} + X^2_{i,2,2} + X^3_{i,1,2} + X^4_{i,1,2} + X^5_{i,1,2,2}$$
$$+X^7_{i,1,2,2} - d_{i,1} - d_{i,2} \qquad \forall i. \qquad (5.242)$$

$$I_{i,t} - B_{i,t} = \sum_{t'''=t''}^{t} \sum_{t''=1}^{t} (X^1_{i,t'',t'''} + X^2_{i,t'',t'''})$$
$$+\sum_{t'''=t''+1}^{t} \sum_{t''=1}^{t-1} (X^3_{i,t'',t'''} + X^4_{i,t'',t'''})$$
$$+\sum_{t''''=t'''}^{t} \sum_{t'''=t''+1}^{t} \sum_{t''=1}^{t-1} (X^5_{i,t'',t''',t''''}$$
$$+X^7_{i,t'',t''',t''''}) + \sum_{t''''=t'''+1}^{t} \sum_{t'''=t''+1}^{t-1} \sum_{t''=1}^{t-2}$$
$$(X^6_{i,t'',t''',t''''} + X^8_{i,t'',t''',t''''}) - \sum_{t''=1}^{t} d_{i,t''} \qquad \forall i \text{ and } t=3,4,..,T. \qquad (5.243)$$

/* Constraints (5.244)–(5.252) represent the boundary conditions */

$$I_{i,T} = B_{i,T} \qquad \forall i. \qquad (5.244)$$

$$\delta^1_{i,i,t} = 0 \qquad \forall i \text{ and } t=2,3,..,T. \qquad (5.245)$$

$$\delta^2_{i,i,t} = 0 \qquad \forall i \text{ and } \forall t. \qquad (5.246)$$

$$\delta^3_{i,i,t} = 0 \qquad \forall i \text{ and } t=1,2,..,T\text{-}1. \qquad (5.247)$$

$$\delta^4_{i,i,t} = 0 \qquad \forall i \text{ and } t=2,3,..,T\text{-}1. \qquad (5.248)$$

$$\delta^5_{i,i,t,t'} = 0 \qquad \forall i,\ t=1,2,..,T\text{-}1 \text{ and } t'=t+1,t+2,..,T. \qquad (5.249)$$

$$\delta^6_{i,i,t,t'} = 0 \qquad \forall i,\ t=1,2,..,T\text{-}2 \text{ and } t'=t+1,t+2,..,T\text{-}1. \qquad (5.250)$$

$$\delta^7_{i,i,t,t'} = 0 \qquad \forall i,\ t=1,2,..,T\text{-}1 \text{ and } t'=t+1,t+2,..,T. \qquad (5.251)$$

$$\delta^8_{i,i,t,t'} = 0 \qquad \forall i,\ t=1,2,..,T\text{-}2 \text{ and } t'=t+1,t+2,..,T\text{-}1. \qquad (5.252)$$

$\delta^1_{0,i,t}$, $\delta^1_{i',i,t}$, $\delta^2_{0,i,t}$, $\delta^2_{i,0,t}$, $\delta^2_{i',i,t}$, $\delta^3_{i',i,t}$, $\delta^4_{0,i,t}$, $\delta^4_{i',i,t}$, $\delta^5_{0,i,t,t'}$, $\delta^5_{i',i,t,t'}$, $\delta^6_{0,i,t,t'}$, $\delta^6_{i',i,t,t'}$, $\delta^7_{i',i,t,t'}$, $\delta^8_{i',i,t,t'}$, $\Delta^1_{i,t,t'}$, $\Delta^2_{i,t,t'}$, $\Delta^3_{i,t,t'}$, $\Delta^4_{i,t,t'}$, $\Delta^5_{i,t,t',t''}$, $\Delta^6_{i,t,t',t''}$, $\Delta^7_{i,t,t',t''}$, $\Delta^8_{i,t,t',t''}$, $\Omega^5_{i,t,t',t''}$, $\Omega^6_{i,t,t',t''}$, $\Omega^7_{i,t,t',t''}$, $\Omega^8_{i,t,t',t''}$, $\alpha_{i,t}$, $\beta_{i,t}$ and $delta_t$ are binary for all i, i', t, t' and t'' as appropriate to the respective binary variable, and all other variables are ≥ 0.

The objective function shown in Eq. (5.109) corresponds to the minimization of sequence-dependent setup costs, holding costs and backorder costs of all products across all time periods. Constraints (5.110)–(5.118) capture a possible complete setup in the beginning of period t, with the production starting in period t. The setup time of the product being set up (using these constraints) should be less than the capacity of that period. Constraints (5.119)–(5.124) correspond to a possible setup occurring anywhere in period t (but not at the beginning or the end of period t), with the production starting in period t. The setup time of the product being set up (using these constraints) should be less than the capacity of period t. Constraints (5.125)–(5.130) correspond to a possible end-of-period setup occurring in period t, with the production starting in period $t + 1$. The setup time of the product being set up (using these constraints) should be less than the capacity of that period. Constraints (5.131)–(5.139) correspond to a possible complete setup done in period t (with the setup started exactly in the beginning of period t, and completed exactly at the end of period t), with the production starting in period $t + 1$. The setup time of the product being set up (using these constraints) should be equal to the capacity of that period. Constraints (5.140)–(5.156) correspond to a possible setup crossover across a number of periods with the production starting in the same period where the setup ends. These constraints hold for the product which has its setup starting in the beginning of period t and ending in the middle of some future period t'. The setup time of the product being set up (using these constraints) should be greater than the capacity of the period in which the setup is initiated. Constraints (5.157)–(5.172) represent a possible setup crossover across a number of periods with the production starting in period $t' + 1$ when the setup ends at the end of period t'. These constraints hold for the product which has its setup starting in the beginning of period t and ends at the end of some future period t'. The setup time of the product being set up (using these constraints) should be greater than the capacity of the period in which the setup is initiated. Constraints (5.173)–(5.186) represent a possible setup crossover across a number of periods with the production starting in the same period where the setup ends. These constraints hold for the product which has its setup starting in the middle of period t and ends in the middle of some future period t'. The setup time of the product being set up (using these constraints) can be greater/less than the capacity of the period in which the setup is initiated. Constraints (5.187)–(5.200) represent a possible setup crossover across a number of periods with the production starting in period $t' + 1$ when the setup ends at the end of period t'. These constraints hold for the product which has its setup starting in the middle of period t and ends at the end of some future period t'. The setup time of the product being set up (using these constraints) should be greater than the capacity of the period in which the setup is initiated. Constraint (5.201) restricts only one product to be setup in the beginning of period 1. Constraint (5.202) indicates that only one product can be tracked with respect to the beginning of every period t. Constraint (5.203) indicates that at most one product can be tracked with respect to the end of every period t. Constraints (5.204)–(5.208) indicate that at most one product can be tracked as the product last setup with respect to the end of every period t. Constraints (5.209)–(5.212) track the product last setup with respect to the beginning of every period t. Constraints (5.213)–(5.215) restrict a product i to be set up in period t with the help of only one type of setup indicator. Constraints (5.216)–(5.218) indicate that only one type of setup can occur between periods t and $t + 1$ for at most one product. When a product i is supposed to be set up, only one of the following types of setup can occur in the border of periods t and $t + 1$, i.e., an end-of period setup in period $t - 1$ using $\delta^3_{i',i,t-1}$; or a beginning of period setup in period t using $\delta^1_{i',i,t}$; or a complete setup in period $t - 1$ using $\delta^4_{i',i,t-1}$; or a complete setup in period t using $\delta^4_{i',i,t}$; or a setup crossover between periods t and $t + 1$ which ends in some period $t' \geq t + 1$ using $\delta^5_{i',i,t,t'}$ or $\delta^6_{i',i,t,t'}$ or $\delta^7_{i',i,t,t'}$ or $\delta^8_{i',i,t,t'}$.

Constraints (5.219)–(5.222) represent the feasibility of production and setup with respect to capacity availability. Constraints (5.223)–(5.226) indicate the flow balance constraints. Constraint (5.227) is the sub-tour elimination constraint. Constraints (5.228)–(5.229) track the product last setup with respect to the beginning of period t. Constraints (5.230)–(5.233) indicate that a production carryover of any product i can take place in period t'' from period t''-2, only if the sum of the production times and setup times of all products in period $t''-1$ is equal to the capacity of $t''-1$. Constraints (5.234)–(5.235) represent the conditions for setting up a product only once in period t. Constraints (5.236)–(5.237) indicate that a production carryover can occur in period t from the preceding period for utmost one product using utmost one type of production carryover indicator. Constraints (5.238)–(5.240) show that if the setup of at least 1 product is initiated in period t (indicated by $delta_t = 1$), then it means that no carryover of either setup or production of any product, with its setup having been initiated in any period up to $t-1$, can be carried over to period $t+1$ and later. This is so because if any product's setup is initiated in period t, then either the production or setup carryover with the setup initiated up to period $t-1$ and carried over to period $t+1$ and later is not feasible. Constraints (5.241)–(5.243) represent the inventory balance constraints. Constraints (5.244)–(5.252) represent the boundary conditions. In Eq. (5.244), $B_{i,T} = 0\ \forall i$ exists because production is not allowed to be carried over to period $T+1$ (i.e. here the demand for all products should be satisfied within period T).

5.5 A Numerical Illustration and Discussion

Let us consider a sample problem instance shown in Table 5.1, with the demand data ($d_{i,t}$) given for ten products across fifteen time periods. The data for the sequence-dependent setup costs ($SC_{\phi,i}$ and $SC_{i',i}$), sequence-dependent setup times ($ST_{\phi,i}$ and $ST_{i',i}$), holding costs (h_i), number of time units required for producing one unit of product i (a_i) and backorder costs (b_i) for all products are also given in the table. For the sake of illustration the proposed models assume equal capacity across different periods. However the proposed models allow for different capacities across different time periods due to the use of C_t, without loss of generality. All these values provided in Table 5.1 are used to illustrate the proposed MM1:CLSP-SD-PCSC and MM2:CLSP-SD-PCSC.

When the proposed Mathematical Model 1 presented in Sect. 5.3.3 (MM1:CLSP-SD-PCSC) is executed for the given sample problem instance given in Table 5.1, a solution (see Table 5.2) with its corresponding Gantt chart shown in Fig. 5.1 is obtained, with the value of the objective function equal to 2202 mu (monetary units). When the proposed Mathematical Model 2 presented in Sect. 5.4.3 (MM2:CLSP-SD-PCSC) is executed for the given sample problem instance given in Table 5.1, a solution (see Table 5.3) with its corresponding Gantt chart shown in Fig. 5.2 is obtained, with the value of the objective function equal to 2202 mu (monetary units). In both cases (Figs. 5.1 and 5.2) it is observed that the production of all products is carried out immediately after setup and there is no interruption or break in production carryover. Hence the solution obtained is feasible and optimal for the problem statement considered with respect to process industries, where sequence-dependent setup costs and setup times are present.

Table 5.1: Product related data

Products	Holding cost (h_i): (mu/period/ unit product carried over)	Backorder cost (b_i): (mu/period/ unit product backordered)	Number of time, units required for producing, one unit of product i (a_i): (time units/unit of product i)
1	2	50,000	1
2	3	50,000	1
3	1	50,000	1
4	4	50,000	1
5	2	50,000	1
6	1	50,000	1
7	2	50,000	1
8	3	50,000	1
9	4	50,000	1
10	1	50,000	1

Demand ($d_{i,t}$): (units)															
	1	2	3	4	5	6	7	8	9	10	11	12	13	14	15
1	0	0	0	0	0	0	0	0	0	0	0	0	0	0	0
2	0	0	0	0	0	0	0	0	0	10	0	0	0	0	0
3	21	0	0	0	0	0	0	20	0	0	0	0	0	0	0
4	0	0	0	0	0	0	0	15	27	0	0	0	0	0	0
5	0	0	0	0	0	40	0	0	0	0	0	0	0	0	0
6	0	0	0	0	0	0	0	0	0	15	0	0	0	0	0
7	0	0	0	0	0	0	0	0	0	0	0	0	0	0	30
8	0	0	0	54	0	0	0	0	0	0	0	0	0	0	0
9	0	0	44	0	0	0	0	0	0	0	0	0	60	0	0
10	0	0	0	0	0	0	0	0	0	0	0	0	0	0	8
Capacity (C_t): (time units)	100	100	100	100	100	100	100	100	100	100	100	100	100	100	100

Sequence-dependent setup cost ($SC_{\phi,i}$ and $SC_{i',i}$): (mu/setup)											
	1	2	3	4	5	6	7	8	9	10	ϕ
1	0	113	79	99	113	147	121	62	101	104	77
2	113	0	76	73	98	75	108	135	166	110	81
3	79	76	0	65	100	109	111	100	136	102	59
4	99	73	65	0	118	98	129	119	159	124	83
5	113	98	100	118	0	121	51	128	137	60	83
6	147	75	109	98	121	0	129	169	200	137	115
7	121	108	111	129	51	129	0	135	140	62	94
8	62	135	100	119	128	169	135	0	86	115	97
9	101	166	136	159	137	200	140	86	0	119	125
10	104	110	102	124	60	137	62	115	119	0	83
ϕ	77	81	59	83	83	115	94	97	125	83	0

Sequence-dependent setup time ($ST_{\phi,i}$ and $ST_{i',i}$): (time units)											
	1	2	3	4	5	6	7	8	9	10	ϕ
1	0	226	158	197	225	294	242	124	202	207	154
2	226	0	151	146	196	149	215	269	332	220	162
3	158	151	0	129	200	218	221	201	272	205	119
4	197	146	129	0	237	196	258	238	317	248	165
5	225	196	200	237	0	242	103	256	275	119	166
6	294	149	218	196	242	0	257	337	399	275	229
7	242	215	221	258	103	257	0	270	280	123	187
8	124	269	201	238	256	337	270	0	171	231	195
9	202	332	272	317	275	399	280	171	0	238	251
10	207	220	205	248	119	275	123	231	238	0	167
ϕ	154	162	119	165	166	229	187	195	251	167	0

Table 5.2: Solution generated by the proposed mathematical model MM1:CLSP-SD-PCSC (corresponding terms in MM1:CLSP-SD-PCSC are used here) for the data provided in Table 5.1 with the corresponding Gantt chart provided in Fig. 5.1

$t = 1$	$t = 2$	$t = 3$	$t = 4$	$t = 5$
$\delta^3_{9,1} = 1$ $\delta^1_{3,1} = 1$ $\Delta^1_{3,1,1} = 1$ $X^1_{3,1,1} = 21$ $s^1_{3,1} = 59$ $s^3_{9,1,1} = 20$ $\Omega^3_{9,1,1} = 1$ $U_{9,1} = 1$ $\xi'_{\phi,3,1} = 1$ $\xi_{3,9,1} = 1$ $ST'_{3,1} = 59$ $ST'_{9,1} = 136$ $del_{\phi,3,1} = 1$ $del_{9,0,1} = 1$	$s^3_{9,1,2} = 100$ $\Omega^3_{9,1,2} = 1$ $del_{9,0,2} = 1$ $del_{0,9,2} = 1$	$\delta^3_{8,3} = 1$ $\Delta^3_{9,1,3} = 1$ $X^3_{9,1,3} = 44$ $s^3_{9,1,3} = 16$ $s^3_{8,3,3} = 40$ $\Omega^3_{8,3,3} = 1$ $\Omega^3_{9,1,3} = 1$ $\xi'_{9,8,3} = 1$ $ST'_{8,3} = 86$ $del_{8,0,3} = 1$ $del_{0,9,3} = 1$ $delta_3 = 1$	$\Delta^3_{8,3,4} = 1$ $X^3_{8,3,4} = 54$ $s^3_{8,3,4} = 46$ $\Omega^3_{8,3,4} = 1$ $del_{8,0,4} = 1$ $del_{0,8,4} = 1$	$\delta^3_{5,5} = 1$ $s^3_{5,5,5} = 68$ $\Omega^3_{5,5,5} = 1$ $\xi'_{8,5,5} = 1$ $ST'_{5,5} = 128$ $del_{5,0,5} = 1$ $del_{0,8,5} = 1$ $delta_5 = 1$
$t = 6$	$t = 7$	$t = 8$	$t = 9$	$t = 10$
$\Delta^3_{5,5,6} = 1$ $X^3_{5,5,6} = 40$ $s^3_{5,5,6} = 60$ $\Omega^3_{5,5,6} = 1$ $del_{5,0,6} = 1$ $del_{0,5,6} = 1$	$\delta^2_{3,7} = 1$ $s^2_{3,7} = 100$ $\xi'_{5,3,7} = 1$ $ST'_{3,7} = 100$ $del_{3,0,7} = 1$ $del_{0,5,7} = 1$ $delta_7 = 1$	$\delta^1_{4,8} = 1$ $\Delta^1_{4,8,8} = 1$ $\Delta^2_{3,7,8} = 1$ $X^2_{3,7,8} = 20$ $X^1_{4,8,8} = 15$ $s^1_{4,8} = 65$ $\xi'_{3,4,8} = 1$ $ST'_{4,8} = 65$ $del_{4,0,8} = 1$ $del_{0,3,8} = 1$ $delta_8 = 1$	$\delta^2_{2,9} = 1$ $\Delta^1_{4,8,9} = 1$ $X^1_{4,8,9} = 27$ $s^2_{2,9} = 73$ $\xi'_{4,2,9} = 1$ $ST'_{2,9} = 73$ $del_{2,0,9} = 1$ $del_{0,4,9} = 1$ $delta_9 = 1$ $U_{7,9} = 9$	$\delta^1_{6,10} = 1$ $\Delta^1_{6,10,10} = 1$ $\Delta^2_{2,9,10} = 1$ $X^2_{2,9,10} = 10$ $X^1_{6,10,10} = 15$ $s^1_{6,10} = 75$ $\xi'_{2,6,10} = 1$ $ST'_{6,10} = 75$ $del_{6,0,10} = 1$ $del_{0,2,10} = 1$ $delta_{10} = 1$

(continued)

Table 5.2: (continued)

$t = 11$	$t = 12$	$t = 13$	$t = 14$	$t = 15$
$\delta^3_{9,11} = 1$	$s^4_{9,11,12} = 100$	$\delta^3_{10,13} = 1$	$s^4_{10,13,14} = 100$	$\delta^1_{7,15} = 1$
$s^4_{9,11,11} = 100$	$\Omega^4_{9,11,12} = 1$	$\Delta^4_{9,11,13} = 1$	$\Omega^4_{10,13,14} = 1$	$\Delta^4_{10,13,15} = 1$
$\Omega^4_{9,11,11} = 1$	$del_{9,0,12} = 1$	$X^4_{9,11,13} = 60$	$del_{10,0,14} = 1$	$\Delta^1_{7,15,15} = 1$
$\xi'_{6,9,11} = 1$	$del_{0,9,12} = 1$	$s^4_{10,13,13} = 19$	$del_{0,10,14} = 1$	$X^1_{7,15,15} = 30$
$ST'_{9,11} = 200$		$\Omega^4_{10,13,13} = 1$		$X^4_{10,13,15} = 8$
$del_{9,0,11} = 1$		$\xi'_{9,10,13} = 1$		$s^1_{7,15} = 62$
$del_{0,6,11} = 1$		$ST'_{10,13} = 119$		$\xi'_{10,7,15} = 1$
$delta_{11} = 1$		$del_{10,0,13} = 1$		$ST'_{7,15} = 62$
		$del_{0,9,13} = 1$		$del_{0,10,15} = 1$
		$delta_{13} = 1$		$del_{7,0,15} = 1$

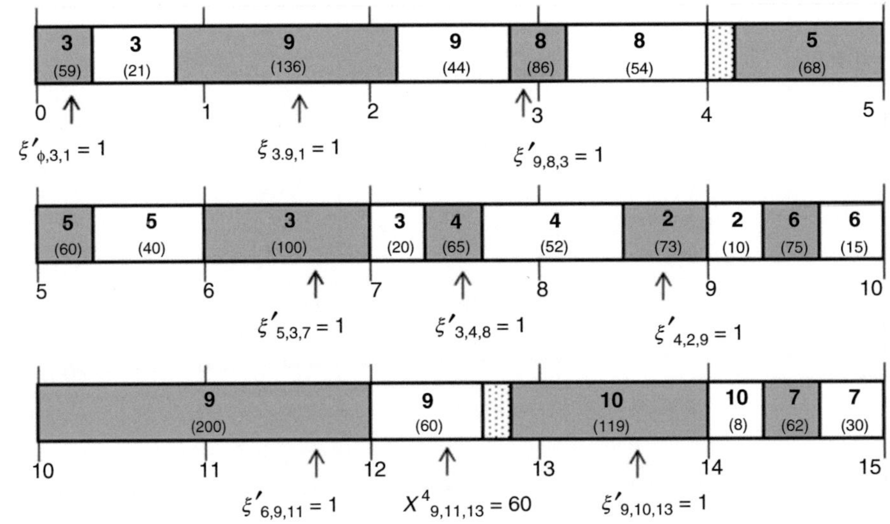

Figure 5.1: Gantt chart for the solution obtained by the proposed model MM1:CLSP-SD-PCSC (for the data given in Table 5.1 and the solution provided in Table 5.2) when sequence-dependent setup costs and setup times are present; $Z = 2202$ mu

Table 5.3: Solution generated by the proposed mathematical model MM2:CLSP-SD-PCSC (corresponding terms in MM2:CLSP-SD-PCSC are used here) for the data provided in Table 5.1 with the corresponding Gantt chart provided in Fig. 5.2

$t=1$	$t=2$	$t=3$	$t=4$	$t=5$
$\delta^1_{0,3,1}=1$ $\delta^7_{3,9,1,3}=1$ $\Delta^1_{3,1,1}=1$ $X^1_{3,1,1}=21$ $s^1_{3,1}=59$ $s^7_{9,1,3,1}=20$ $\Omega^7_{9,1,3,1}=1$ $a_{3,1}=1$	$s^7_{9,1,3,2}=100$ $\Omega^7_{9,1,3,2}=1$ $a_{9,2}=1$	$\delta^7_{9,8,3,4}=1$ $\Delta^7_{9,1,3,3}=1$ $X^3_{9,1,3,3}=44$ $s^7_{9,1,3,3}=16$ $s^7_{8,3,4,3}=40$ $\Omega^7_{9,1,3,3}=1$ $\Omega^7_{8,3,4,3}=1$ $a_{9,3}=1$ $delta_3=1$	$\Delta^7_{8,3,4,4}=1$ $X^7_{8,3,4,4}=54$ $s^7_{8,3,4,4}=46$ $\Omega^7_{8,3,4,4}=1$ $a_{8,4}=1$ $\beta_{8,4}=1$	$\delta^7_{8,5,5,6}=1$ $s^7_{5,5,6,5}=99.999$ $\Omega^7_{5,5,6,5}=1$ $a_{8,5}=1$ $delta_5=1$
$t=6$	$t=7$	$t=8$	$t=9$	$t=10$
$\Delta^7_{5,5,6,6}=1$ $X^7_{5,5,6,6}=40$ $s^7_{5,5,6,6}=28.001$ $\Omega^7_{5,5,6,6}=1$ $a_{5,6}=1$	$\delta^4_{5,3,7}=1$ $s^4_{3,7}=100$ $a_{3,7}=1$ $delta_7=1$	$\delta^2_{3,4,8}=1$ $\Delta^2_{4,8,8}=1$ $\Delta^4_{3,7,8}=1$ $X^4_{3,7,8}=20$ $X^2_{4,8,8}=15$ $s^2_{4,8}=65$ $U_{4,8}=1$ $a_{3,8}=1$ $delta_8=1$	$\delta^3_{4,2,9}=1$ $\Delta^2_{4,8,9}=1$ $X^2_{4,8,9}=27$ $s^3_{2,9}=73$ $a_{4,9}=1$ $delta_9=1$	$\delta^2_{2,6,10}=1$ $\Delta^2_{6,10,10}=1$ $\Delta^3_{2,9,10}=1$ $X^3_{2,9,10}=10$ $X^2_{6,10,10}=15$ $s^2_{6,10}=75$ $U_{6,10}=1$ $a_{2,10}=1$ $delta_{10}=1$
$t=11$	$t=12$	$t=13$	$t=14$	$t=15$
$\delta^6_{6,9,11,12}=1$ $s^7_{9,11,12,11}=100$ $\Omega^6_{9,11,12,11}=1$ $a_{6,11}=1$ $delta_{11}=1$	$s^7_{9,11,13,12}=100$ $\Omega^6_{9,11,12,12}=1$ $s^5_{9,11,12,12}=100$ $a_{9,12}$	$\delta^8_{9,10,13,14}=1$ $\Delta^6_{9,11,12,13}=1$ $X^6_{9,11,12,13}=60$ $s^7_{9,11,13,13}=0.001$ $s^8_{10,13,14,13}=19$ $\Omega^8_{10,13,14,13}=1$ $a_{9,13}=1$ $delta_{13}=1$	$s^8_{10,13,14,14}=100$ $\Omega^8_{10,13,14,14}=1$ $a_{10,14}=1$	$\delta^2_{10,7,15}=1$ $\Delta^8_{10,13,14,15}=1$ $\Delta^2_{7,15,15}=1$ $X^2_{7,15,15}=30$ $X^8_{10,13,14,15}=8$ $s^2_{7,15}=62$ $U_{7,15}=1$ $a_{10,15}=1$

Note: The binary variable $a_{7,16}=1$ also gets generated to satisfy the Constraint (5.226). However, the value does not have any significance with respect to the output of the model.

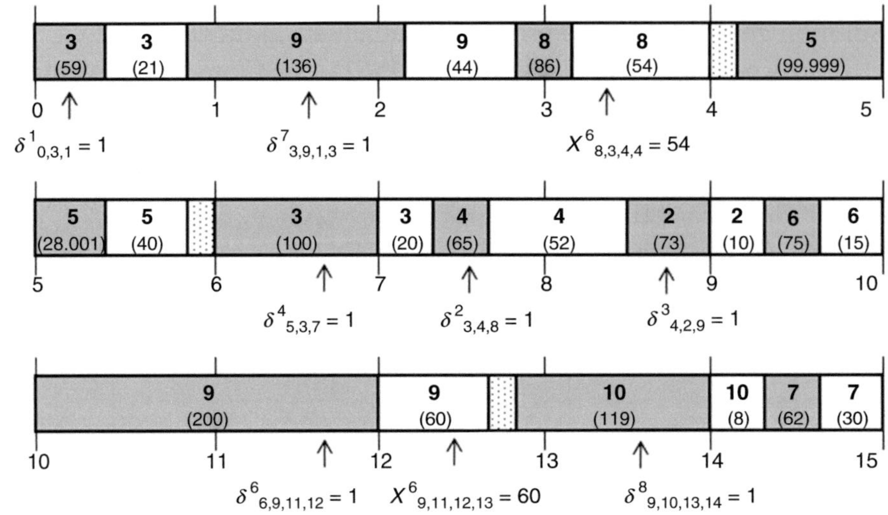

Figure 5.2: Gantt chart for the solution obtained by the proposed model MM2:CLSP-SD-PCSC (for the data given in Table 5.1 and the solution provided in Table 5.3) when sequence-dependent setup costs and setup times are present; $Z = 2202$ mu

5.6 Computational Experience

In this section, the solution times of the proposed MM1:CLSP-SD-PCSC and MM2:CLSP-SD-PCSC are compared. In Table 5.4, the solution time (to obtain an optimal solution) obtained by executing the generalized MM1:CLSP-SD-PCSC and MM2:CLSP-SD-PCSC for each of the 9 sample problem instances of various product and time-period combinations is presented. All the problem instances have been executed on a Pentium 3.10 GHz Windows 7 workstation with 4.00 GB RAM, using CPLEX v12.4. It is believed that if the same problem instances are run on the latest version of CPLEX (v12.6.3), the solution times will improve. While using CPLEX, the values of the mixed integer optimality gap tolerance and integrality tolerance were set to zero, in order to obtain accuracy in results and maintain the integrality with respect to binary variables. Alternate optimal solutions are obtained for the problem instances when the default tolerance values of the CPLEX are retained.

The two tolerances which had been set to zero instead of using the default values were the integrality tolerance and the MIP tolerance. The parameter integrality tolerance sets the amount by which a computed solution value for an integer variable can violate integrality; it does not specify an amount by which CPLEX relaxes integrality. In order to avoid violation of integrality and to obtain the exact values without floating the integer solution values, it has been set it exactly to zero such that the resulting solution values are exact and rounding off the solution values need not be done manually. Sometimes having tolerance set to zero helps to compare our results with other exact methods (other mathematical models addressing the same problem), so that no variables which need not necessarily float are present in the final solution. When a problem instance considering 5 products and 5 periods (see Table 5.1) was executed with default tolerance setting the value of an integer variable which was supposed to provide a value 0 gave a value 0.001 which resulted in a higher value of objective function (Z= 790.001 mu). This variable took a value zero when the tolerance was set to zero, resulting in an optimal value of objective function (Z= 790 mu). Therefore, in order to obtain the exact values of the variables, the tolerance value is set to 0. The parameter MIP tolerance is the gap between the best integer objective and the objective of the best node remaining. When this difference falls below the value of this parameter, the mixed integer optimization is stopped. In order to stop the optimization only at the best integer objective, the parameter is set to zero. The following settings are assumed while performing the computations: The sequence-dependent setup costs (mu/setup) of the products range from 22 mu/setup to 428 mu/setup; the holding costs (mu/period/unit product carried over) of the products range from 1 mu/period/unit product carried over to 4 mu/period/unit product carried over; the sequence-dependent setup time (time units) of the products (per setup) ranges from 11 time units to 214 time units and the backorder cost (mu/period/unit product backordered) of all products is assumed to be 50,000 mu/period/unit product backordered (which implies no backorder). The following settings are assumed while performing the computations: The sequence-dependent setup costs (mu/setup) of the products range from 22 mu/setup to 428 mu/setup; the holding costs (mu/period/unit product carried over) of the products range from 1 mu/period/unit product carried over to 4 mu/period/unit product carried over; the sequence-dependent setup time (time units) of the products (per setup) range from 11 time units to 214 time units and the backorder cost (mu/period/unit product backordered) of all products is assumed to be 50,000 mu/period/unit product backordered (which implies no backorder). Throughout this work 100 time units correspond to one time period (t) and the number of time units required for producing one unit of product i (a_i) is equal to 1 time unit per unit of product i.

Table 5.4: Computational time (in sec.) for various problem instances

Mathematical models	No. of time periods / No. of products	5	10	15
MM1:CLSP-SD-PCSC	5	0.78	35.07	14.84
MM2:CLSP-SD-PCSC		0.56	1.83	9.34
MM1:CLSP-SD-PCSC	10	1.64	27071.82	3462.12
MM2:CLSP-SD-PCSC		2.29	4520.86	425.18
MM1:CLSP-SD-PCSC	15	15.94	–	–
MM2:CLSP-SD-PCSC		6.35	26311.75	–

Note:
(a) When the problem instance with respect to MM1:CLSP-SD-PCSC is executed assuming 15 products and 10 periods, the solver fails with a file error after 14 h due to insufficient space in the hard disk.
(b) When the problem instance with respect to MM1:CLSP-SD-PCSC is executed assuming 15 products and 15 periods, the solver fails with a file error after 36.6 h due to insufficient space in the hard disk.
(c) When the problem instance with respect to MM2:CLSP-SD-PCSC is executed assuming 15 products and 15 periods, the solver fails with a file error after 8.5 h due to insufficient space in the hard disk.

5.7 Summary

In this chapter, the capacitated lot sizing problem with sequence-dependent setup times and setup costs, and with production carryover and setup crossover across periods is addressed for process industries by proposing two mathematical models. These models address real-life scenarios present in most process industries which require the production of a product to occur immediately after the product is set up and continuously produced without any interruption, preventing the presence of idle time during the course of production carryover of a product. The models assume the setup cost and holding cost of a product to be time independent, and sequence-dependent setup costs and setup times. The CLSP belongs to a set of problems that are called NP-hard. Therefore, it is very difficult to optimally solve large instances of this problem involving sequence-dependent setups due to which heuristics can be proposed in future to solve this problem. From the two exact approaches proposed, MM2:CLSP-SD-PCSC solves faster than MM1:CLSP-SD-PCSC. For large problem instances, heuristics can be developed similar to Chap. 4 for the CLSP with sequence-dependent costs and times.

CHAPTER 6

Summary Concerning Theoretical Developments

Lot sizing is a major decision taken during the planning of production of various products in process and manufacturing industries. The lot sizing problems can be classified into continuous lot sizing problem (economic lot scheduling problem) and dynamic lot sizing problem. The time scale considered is continuous and infinite in the continuous lot sizing problem, whereas a discrete time scale is considered in dynamic lot sizing problems. The dynamic lot sizing problems are further classified into uncapacitated and capacitated lot sizing problems based on their capacity restrictions. The capacitated lot sizing problems are further classified into small bucket and big bucket lot sizing models depending upon the number of setups that are allowed in a given time period. The discrete lot sizing and scheduling problem (DLSP), continuous setup lot sizing problem (CSLP) and the proportional lot sizing and scheduling problem (PLSP) come under the small bucket lot sizing models, and the capacitated lot sizing problem (CLSP) comes under the big bucket lot sizing model.

The CLSP (capacitated lot sizing problem), which is one of the basic lot sizing models studied extensively in literature, is the main focus of this book. This big bucket lot sizing model considers a single machine in which multiple products can be produced in every period, under capacity limitations. The products are made up of a single level and face a deterministic demand. The demand for the products considered is given over a finite time horizon, and the time units considered are discrete (such as a shift/a day/a week, etc.). When a product is setup it consumes a time for its setup. The capacity of the machine is assumed in time units and is consumed by the setup time and production time of the products. The problem deals with the minimization of the total costs involved in lot sizing, thereby planning the production of several products over a number of periods, satisfying all the demand requirements and without exceeding the capacity limitations. Excess quantity of a product can also be produced and stored in a period, to satisfy the demand of a future period, provided capacity is available to carry out the production. Earlier research attempts in CLSP focused on the inclusion of setup costs, setup times, and production carryover of a product while developing mathematical models for the CLSP. Production carryover is a phenomenon where the state of a setup of a product in period can be carried forward to the next period without any interruption of setup across consecutive periods, thereby avoiding more than one setup of the same product in the consecutive periods, in order to reduce the cost for setup and also utilize the capacity efficiently. Recent literature in CLSP addressed the possibility of a split setup and setup crossover across a number of periods in the proposed mathematical models, thereby attempting to utilize the capacity of the machine efficiently. Setup splitting refers to the initiation of setup of a product i in period t and completion of setup of the product in the next period $t + 1$. Setup crossover is a phenomena where the setup started in period t can be carried over to period t', followed by

R. Ramya et al., *Capacitated Lot Sizing Problems in Process Industries*, https://doi.org/10.1007/978-3-030-01222-9_6

its production in time period t' or $t'' > t'$, depending upon where the setup of product i ends (here t' can be any one of the periods $t + 1 \leq t' \leq T$).

This book deals with a class of capacitated lot sizing problems, addressing various situations present in process industries. Mathematical models and heuristics have been developed to address the various class of lot sizing problems. The mathematical models assume the presence of setup costs, holding costs, backorder costs, setup times, setup crossover and production carryover. Sequence-dependent setup costs and setup times are also considered in one of the chapters. The models and the heuristic developed can be applied to various process industries to determine the quantity of products needed to be produced, given the demand requirements for various products across a finite time horizon, thereby maintaining the feasibility of the production and setup with respect to the capacity availability; also taking into consideration, most of the real-life situations present in these industries such as production immediately after setup and uninterrupted production carryover across periods. The main objective considered in this class of capacitated lot sizing problems is the minimization of the setup cost, holding cost and backorder cost of all products across all time periods.

In the introduction presented in Chap. 1, the history of the production planning problem has been presented starting from the earliest models in lot sizing: Harris EOQ (economic order quantity) formula, Wilsons's (Q,r) model and the dynamic lot sizing model proposed by Wagner and Whitin. Following this, the basic characteristics and attributes of lot sizing models have been stated based on time and product. Some of the time based characteristics and attributes of the lot sizing models which have been explained are the planning horizon, time scale, parameters/data, objective function, cost components, capacity, number of resources and setup operation. The product based characteristics and attributes which have been explained are number of products, inventory restrictions and service policy. Following this, various classifications of lot sizing models have been presented. They are: (1) Continuous lot sizing problem: economic lot scheduling problem; and the (2) dynamic lot sizing problem. The dynamic lot sizing problem has been classified into (a) Uncapacitated lot sizing problem: Wagner and Whitin dynamic lot sizing problem, and (b) Capacitated lot sizing problem. The capacitated lot sizing problems are further categorized into (i) Small bucket lot sizing models and (ii) Big bucket lot sizing models. The discrete lot sizing and scheduling problem (DLSP), continuous setup lot sizing problem (CSLP) and the proportional lot sizing and scheduling problem (PLSP) come under the small bucket lot sizing models and the capacitated lot sizing problem (CLSP) comes under the big bucket lot sizing model. Following this, the overall CLSP literature has been classified into: (a) CLSP without production carryover across periods and without sequence-dependent setups; (b) CLSP without production carryover across periods and with sequence-dependent setups ; (c) CLSP with production carryover across periods and without sequence-dependent setups; and (d) CLSP with production carryover across periods and with sequence-dependent setups, followed by its discussion. The scenario of production of a product starting immediately after its setup completion and no interruption of its production once started is encountered in process industries. These aspects are not addressed/considered in the existing literature in the existing models in the context of process industries especially with respect to sequence dependent setups.

In Chap. 2, the production planning in discrete manufacturing industries and continuous manufacturing industries has been presented with the help of various examples. A real-life case study is discussed. This real-life case study and the gaps identified in literature (see Belo-Filho et al. (2013)) form the motivation for the models and heuristics proposed in this book.

In Chap. 3, the introduction and problem definition of the CLSP with production carryover and setup crossover across periods have been presented. Long setup times are prevalent in almost all process industries. In industries where there are long setup times, manufacturers allow the setup to be carried over across periods. Allowing production carryover, setup splitting and setup crossover across periods help the manufacturer to use the capacity effectively. Therefore, there can be three ways by which a machine is set up for producing a product when long setup times are present. They are: (i) a machine is completely setup for product i anywhere in period t, and the production starts in period t itself and the production may be continued to the period(s) thereafter (this aspect is referred to as production carryover in this work); (ii) when there is enough capacity left at the end of a period, it can be utilized in making a complete setup for a product, followed by its production in time period $t + 1$ (this aspect is referred to as end-of-period setup in this work). An end-of-period setup of product i can occur when an amount of capacity is left at the end of period t to set up the product, or it can also refer to the setup which takes a value equal to the capacity of the period in which its setup is initiated; and (iii) the setup started in period t can be carried over to period t', followed by its production in time period t' or $t'' > t'$ depending upon where the setup of product i ends (this aspect is referred to as setup crossover in this work). The phenomenon setup crossover also includes the concept of setup splitting between periods t and $t + 1$, when the setup time of product i is less than the capacity of period t in which its setup is initiated, and less than the capacity of period $t + 1$ in which its setup is completed. In process industries, the production should start immediately after the setup, and no break during the course of production of a product is allowed. A brief description of these aspects present in some process industries are also explained in this chapter with the help of some examples. Therefore, in order to address situations arising across such process industries, and hence to solve the CLSP in those industries, i.e., production to commence immediately after setup, and no break or discontinuity during the course of production, along with the presence of long setup times, a generalized mathematical model and a heuristic with respect to the mathematical model has been proposed in this chapter. Special cases of the mathematical models considering time-dependent and time-independent cost structure have also been presented. Belo-Filho et al. (2013) had already worked on the CLSP with setup crossover and production carryover across periods, allowing the presence of backorders. These authors considered the presence of long setup times. However, they did not address real-life situations prevailing in most process industries such as production commencement immediately after the completion of setup and no break or discontinuity during the course of production. Numerical illustrations and discussion have been presented for the mathematical model and its special cases (Gantt charts corresponding to the solution obtained for the numerical illustrations have also been presented). The observations in the model by Belo-Filho et al. (2013) have also been presented, and their results have been compared with the proposed mathematical models. From the results it can be observed that this existing model does not consider real-life situations present in process industries such as production immediately after setup and uninterrupted production carryover. A heuristic with respect to the mathematical model has been proposed. The heuristic adapts the idea of "Relax-and-Fix" heuristic proposed by some authors in literature, but has its own novelty in terms of determining the size of the time window and solving each sub-problem. The computational time for generalized MM1:CLSP-PCSC and the computational time of the heuristic with respect to MM1:CLSP-PCSC are also presented.

In Chap. 4, a second mathematical model has been proposed for the CLSP-PCSC along with a comprehensive heuristic with respect to the mathematical model. This mathematical model and heuristic also address the presence of long setup times, and real-life situations prevailing in most process industries such as production commencement immediately after the completion of setup and no break or discontinuity during the course of production. Numerical illustrations and discussion have been presented for the mathematical model and its special cases (Gantt charts corresponding to the solution obtained for the numerical illustrations have also been presented). A comprehensive heuristic is also proposed based on MM2:CLSP-PCSC. The model MM2:CLSP-PCSC is different from MM1:CLSP-PCSC and the heuristic is comprehensive i.e., in the heuristic with respect to MM1:CLSP-PCSC only one left shift and right shift of setup/production ahead of or after the current time period is allowed and can be easily applied when equal capacity is present across periods. However, since the extension of heuristic of MM1:CLSP-PCSC is tedious when the capacity is unequal across periods, the heuristic based on MM2:CLSP-PCSC is proposed. In the heuristic with respect to MM2:CLSP-PCSC the left shift/right shift can be made across the entire horizon. The computational time for generalized MM2:CLSP-PCSC is also presented.

In Chap. 5, the capacitated lot sizing problem with sequence-dependent setup times and setup costs and with production carryover and setup crossover across periods is addressed for process industries by developing two mathematical models (MM1:CLSP-SD-PCSC and MM2:CLSP-SD-PCSC). In process industries, products may have short or long setup times, and the setup cost and setup time may be sequence independent or sequence dependent. Sequence-dependent setups exist in industries where the setup time of a machine not only depends on the time to set up the particular product, it can also depend upon the previous product for which the machine was set up for, and other procedures which need to be undertaken before setting up the next product. Considering i' and i as two products which need to be set up on a single machine successively, when the setup time of a product on the machine depends on the time taken to set up product i after setting up the previous product i', the setup is called as sequence-dependent setup and the corresponding time taken for setting up the product is called as sequence-dependent setup time. The corresponding cost involved in setting up product i after setting up product i' is called sequence-dependent setup cost. The basic assumptions have been presented followed by the presentation of both the mathematical models. The proposed models mainly address real-life situations present in most process industries where the presence of idle time is not allowed between the setup of a product and its consecutive production, and when the production carryover of the product takes place across periods. Along with this, they address the presence of long setup times and sequence-dependent setup times and setup costs. A numerical illustration and discussion on both the mathematical models has been presented with the help of a numerical example. The computational time for several product-time period combinations has also been presented.

The theoretical work can be extended by developing meta-heuristics or Lagrangian relaxation based heuristics to address the same problem. The mathematical models can also be extended to include parallel machines or multiple machines, and multiple levels of a product can also be included. Lost sales can be introduced into the proposed models. Stochastic variations in demand can be yet another aspect for future research work.

REFERENCES

Abdelmaguid, T. F. 2015. A neighborhood search function for flexible job shop scheduling with separable sequence-dependent setup times. *Applied Mathematics and Computation* 260(1): 188–203.

Aggarwal, A., and J. K. Park. 1993. Improved algorithms for economic lot size problems. *Operations research* 41(3): 549–571.

Akartunalı, K., and A. J. Miller. 2009. A heuristic approach for big bucket multi-level production planning problems. *European Journal of Operational Research* 193(2): 396–411.

Almada-Lobo, B., and R. J. James. 2010. Neighbourhood search meta-heuristics for capacitated lot-sizing with sequence-dependent setups. *International Journal of Production Research* 48(3): 861–878.

Almada-Lobo, B., D. Klabjan, M. Antónia carravilla, and J. F. Oliveira. 2007. Single machine multi-product capacitated lot sizing with sequence-dependent setups. *International Journal of Production Research* 45(20): 4873–4894.

Almada-Lobo, B., J. F. Oliveira, and M. A. Carravilla. 2008. Production planning and scheduling in the glass container industry: A vns approach. *International Journal of Production Economics* 114(1): 363–375.

Amorim, P., H.-O. Günther, and B. Almada-Lobo. 2012. Multi-objective integrated production and distribution planning of perishable products. *International Journal of Production Economics* 138(1): 89–101.

Bahl, H. C. 1983. Column generation based heuristic algorithm for multi-item scheduling. *AIIE Transactions* 15(2): 136–141.

Bahl, H. C., L. P. Ritzman, and J. N. Gupta. 1987. Or practice—determining lot sizes and resource requirements: A review. *Operations Research* 35(3): 329–345.

Belo-Filho, M. A., F. M. Toledo, and B. Almada-Lobo. 2013. Models for capacitated lot-sizing problem with backlogging, setup carryover and crossover. *Journal of the Operational Research Society* 65(11): 1735–1747.

Belvaux, G. and L. A. Wolsey. 2000. Bc—prod: a specialized branch-and-cut system for lot-sizing problems. *Management Science* 46(5): 724–738.

Belvaux, G. and L. A. Wolsey. 2001. Modelling practical lot-sizing problems as mixed-integer programs. *Management Science* 47(7): 993–1007.

R. Ramya et al., *Capacitated Lot Sizing Problems in Process Industries*, https://doi.org/10.1007/978-3-030-01222-9

Berndt, E. R. 2002. Pharmaceuticals in us health care: determinants of quantity and price. *Journal of Economic Perspectives* 16(4): 45–66.

Bilde, O. and J. Krarup. 1977. Sharp lower bounds and efficient algorithms for the simple plant location problem. *Annals of Discrete Mathematics* 1: 79–97.

Billington, P. J., J. O. McClain, and L. J. Thomas. 1983. Mathematical programming approaches to capacity-constrained mrp systems: review, formulation and problem reduction. *Management Science* 29(10): 1126–1141.

Bitran, G. R., T. L. Magnanti, and H. H. Yanasse. 1984. Approximation methods for the uncapacitated dynamic lot size problem. *Management Science* 30(9): 1121–1140.

Bitran, G. R. and H. H. Yanasse. 1982. Computational complexity of the capacitated lot size problem. *Management Science* 28(10): 1174–1186.

Briskorn, D. 2006. A note on capacitated lot sizing with setup carry over. *IIE Transactions* 38(11): 1045–1047.

Brown, G., J. Keegan, B. Vigus, and K. Wood. 2001. The kellogg company optimizes production, inventory, and distribution. *Interfaces* 31(6): 1–15.

Camargo, V. C., F. M. B. Toledo, and B. Almada-Lobo. 2012. Three time-based scale formulations for the two-stage lot sizing and scheduling in process industries. *Journal of the Operational Research Society* 63(11): 1613–1630.

Caserta, M., A. Ramirez, and S. Voß. 2010. A math-heuristic for the multi-level capacitated lot sizing problem with carryover. *Applications of evolutionary computation*, 462–471. Springer.

Caserta, M., A. Ramirez, S. Voß, and R. Moreno. 2009. A hybrid algorithm for the multi level capacitated lot sizing problem with setup carry-over. *Logistik management*, 123–138. Springer.

Caserta, M. and S. Voß. 2013. A mip-based framework and its application on a lot sizing problem with setup carryover. *Journal of Heuristics* 19(2): 295–316.

Clark, A., M. Mahdieh, and S. Rangel. 2014. Production lot sizing and scheduling with non-triangular sequence-dependent setup times. *International Journal of Production Research* 52(8): 2490–2503.

Clark, A. R. and S. J. Clark. 2000. Rolling-horizon lot-sizing when set-up times are sequence-dependent. *International Journal of Production Research* 38(10): 2287–2307.

Diaby, M., H. C. Bahl, M. H. Karwan, and S. Zionts. 1992. A lagrangean relaxation approach for very-large-scale capacitated lot-sizing. *Management Science* 38(9): 1329–1340.

Dillenberger, C., L. F. Escudero, A. Wollensak, and W. Zhang. 1993. On solving a large-scale resource allocation problem in production planning. *Operations research in production planning and control*, 105–119. Springer.

Dillenberger, C., L. F. Escudero, A. Wollensak, and W. Zhang. 1994. On practical resource allocation for production planning and scheduling with period overlapping setups. *European Journal of Operational Research* 75(2): 275–286.

Dogramaci, A., J. C. Panayiotopoulos, and N. R. Adam. 1981. The dynamic lot-sizing problem for multiple items under limited capacity. *AIIE Transactions* 13(4): 294–303.

Drexl, A. and K. Haase. 1995. Proportional lotsizing and scheduling. *International Journal of Production Economics* 40(1): 73–87.

Eisenhut, P. 1975. A dynamic lot sizing algorithm with capacity constraints. *AIIE Transactions* 7(2): 170–176.

Elmaghraby, S. E. 1978. The economic lot scheduling problem (elsp): review and extensions. *Management Science* 24(6): 587–598.

Eppen, G. D., F. Gould, and B. P. Pashigian. 1969. Extensions of the planning horizon theorem in the dynamic lot size model. *Management Science* 15(5): 268–277.

Eppen, G. D. and R. K. Martin. 1987. Solving multi-item capacitated lot-sizing problems using variable redefinition. *Operations Research* 35(6): 832–848.

Eren, T. and E. Güner. 2006. A bicriteria scheduling with sequence-dependent setup times. *Applied Mathematics and Computation* 179(1): 378–385.

Federgruen, A. and M. Tzur. 1991. A simple forward algorithm to solve general dynamic lot sizing models with n periods in 0 (n log n) or 0 (n) time. *Management Science* 37(8): 909–925.

Ferreira, D., A. R. Clark, B. Almada-Lobo, and R. Morabito. 2012. Single-stage formulations for synchronised two-stage lot sizing and scheduling in soft drink production. *International Journal of Production Economics* 136(2): 255–265.

Fleischmann, B. 1990. The discrete lot-sizing and scheduling problem. *European Journal of Operational Research* 44(3): 337–348.

Florian, M., J. K. Lenstra, and A. Rinnooy Kan. 1980. Deterministic production planning: Algorithms and complexity. *Management science* 26(7): 669–679.

Fransoo, J. C. 1993. *Production control and demand management in capacitated flow process industries*. Technische Universiteit, 1993.

Gao, Y. 2000. A heuristic procedure for the capacitated lot sizing problem with set-up carry-over. *Control Theory and Applications* 17(6): 937–940.

Ganesh, K., S. Mohapatra, and S. Nagarajan. 2014. *Design and development of knowledge management for manufacturing: framework, solution and strategy*, Springer.

Gnoni, M., R. Iavagnilio, G. Mossa, G. Mummolo, and A. Di Leva. 2003. Production planning of a multi-site manufacturing system by hybrid modelling: A case study from the automotive industry. *International Journal of Production Economics* 85(2): 251–262.

Gopalakrishnan, M. 2000. A modified framework for modelling set-up carryover in the capacitated lotsizing problem. *International Journal of Production Research* 38(14): 3421–3424.

Gopalakrishnan, M., K. Ding, J.-M. Bourjolly, and S. Mohan. 2001. A tabu-search heuristic for the capacitated lot-sizing problem with set-up carryover. *Management Science* 47(6): 851–863.

Gopalakrishnan, M., D. Miller, and C. Schmidt. 1995. A framework for modelling setup carryover in the capacitated lot sizing problem. *International Journal of Production Research* 33(7): 1973–1988.

Goren, H. G., S. Tunali, and R. Jans. 2012. A hybrid approach for the capacitated lot sizing problem with setup carryover. *International Journal of Production Research* 50(6): 1582–1597.

Grünert, T. 1998. *Multi-level sequence dependent dynamic lotsizing and scheduling*. Shaker Verlag.

Guimarães, L., D. Klabjan, and B. Almada-Lobo. 2014. Modeling lotsizing and scheduling problems with sequence dependent setups. *European Journal of Operational Research* 239(3): 644–662.

Haase, K. 1996. Capacitated lot-sizing with sequence dependent setup costs. *Operations-Research-Spektrum* 18(1): 51–59.

Haase, K. 1998. *Capacitated lot-sizing with linked production quantities of adjacent periods*. Springer.

Haase, K. and A. Drexl. 1994. Capacitated lot-sizing with linked production quantities of adjacent periods. *Operations Research'93*, 212–215. Springer.

Haase, K. and A. Kimms. 2000. Lot sizing and scheduling with sequence-dependent setup costs and times and efficient rescheduling opportunities. *International Journal of Production Economics* 66(2): 159–169.

Harris, F. W. 1990. How many parts to make at once. *Operations Research* 38(6): 947–950.

Hung, Y.-F. and K.-L. Chien. 2000. A multi-class multi-level capacitated lot sizing model. *Journal of the Operational Research Society* 51(11): 1309–1318.

James, R. J. and B. Almada-Lobo. 2011. Single and parallel machine capacitated lotsizing and scheduling: New iterative mip-based neighborhood search heuristics. *Computers & Operations Research* 38(12): 1816–1825.

Jeong, B. and Y.-D. Kim. 2014. Minimizing total tardiness in a two-machine re-entrant flowshop with sequence-dependent setup times. *Computers & Operations Research* 47: 72–80.

Kang, S., K. Malik, and L. J. Thomas. 1999. Lotsizing and scheduling on parallel machines with sequence-dependent setup costs. *Management Science* 45(2): 273–289.

Karimi, B., S. Ghomi, and J. Wilson. 2006. A tabu search heuristic for solving the clsp with backlogging and set-up carry-over. *Journal of the Operational Research Society* 57(2): 140–147.

Karimi, B., S. F. Ghomi, and J. Wilson. 2003. The capacitated lot sizing problem: a review of models and algorithms. *Omega* 31(5): 365–378.

Karmarkar, U. S. and L. Schrage. 1985. The deterministic dynamic product cycling problem. *Operations Research* 33(2): 326–345.

Katok, E., H. S. Lewis, and T. P. Harrison. 1998. Lot sizing in general assembly systems with setup costs, setup times, and multiple constrained resources. *Management Science* 44(6): 859–877.

Kimms, A. 1997. *Multi-level lot sizing and scheduling*. Heidelberg: Physica-Verlag.

Kimms, A. and A. Drexl. 1998. Proportional lot sizing and scheduling: Some extensions. *Networks* 32(2): 85–101.

Kovács, A., K. N. Brown, and S. A. Tarim. 2009. An efficient mip model for the capacitated lot-sizing and scheduling problem with sequence-dependent setups. *International Journal of Production Economics* 118(1): 282–291.

Kreipl, S. and M. Pinedo. 2004. Planning and scheduling in supply chains: an overview of issues in practice. *Production and Operations management* 13(1): 77–92.

Kwak, I.-S. and I.-J. Jeong. 2011. A hierarchical approach for the capacitated lot-sizing and scheduling problem with a special structure of sequence-dependent setups. *International Journal of Production Research* 49(24): 7425–7439.

Lambrecht, M. and H. Vanderveken. 1979. Heuristic procedures for the single operation, multi-item loading problem. *AIIE Transactions* 11(4): 319–326.

Lang, J. C. and Z.-J. M. Shen. 2011. Fix-and-optimize heuristics for capacitated lot-sizing with sequence-dependent setups and substitutions. *European Journal of Operational Research* 214(3): 595–605.

Maes, J., J. O. McClain, and L. N. van Wassenhove. 1991. Multilevel capacitated lotsizing complexity and lp-based heuristics. *European Journal of Operational Research* 53(2): 131–148.

Magnanti, T. L. and R. Vachani. 1990. A strong cutting plane algorithm for production scheduling with changeover costs. *Operations Research* 38(3): 456–473.

Magnusson, T. 2001. *The capacitated lot-sizing and scheduling problem with sequence-dependent setup costs and setup times*. Ph.D. thesis, University of Minnesota.

Manne, A. S. 1958. Programming of economic lot sizes. *Management Science* 4(2): 115–135.

Millar, H. H. and M. Yang. 1993. An application of lagrangean decomposition to the capacitated multi-item lot sizing problem. *Computers & Operations Research* 20(4): 409–420.

Millar, H. H. and M. Yang. 1994. Lagrangian heuristics for the capacitated multi-item lot-sizing problem with backordering. *International Journal of Production Economics* 34(1): 1–15.

Mohammadi, M., S. F. Ghomi, B. Karimi, and S. A. Torabi. 2010. Rolling-horizon and fix-and-relax heuristics for the multi-product multi-level capacitated lotsizing problem with sequence-dependent setups. *Journal of Intelligent Manufacturing* 21(4): 501–510.

Mohan, S., M. Gopalakrishnan, R. Marathe, and A. Rajan. 2012. A note on modelling the capacitated lot-sizing problem with set-up carryover and set-up splitting. *International Journal of Production Research* 50(19): 5538–5543.

Mohapatra, S. 2009. Better Healthcare at reduced cost through electronic integration of patient care data. *International Journal of Electronic Healthcare* 5(1): 27–54.

Mohapatra, S. 2010. Improvised process for quality through quantitative project management: an experience from software development projects. *International Journal of Information and Communication Technology (IJICT)* 2(4): 34–56.

Mohapatra, S. 2015a. Business school education and technology–a case study. *Education and Information Technologies* 20(2): 335–346.

Mohapatra, S. 2015b. Requirement management – controlling quality at the upstream in commercial software project management. *International Journal of Applied Engineering Research* 10(3): 8203–8219.

Mohapatra, S. 2015c. Using integrated information system for patient benefits, a case study in India. *International Journal of Healthcare Management* 8(4): 262–271.

Mohapatra, S. 2017. An empirical study for finding factors that would optimise productivity and quality in IT business. *International Journal of Productivity and Quality Management* 20(2): 169–196.

Mohapatra, S. and J. Anuj. 2010. Green IT framework for small and medium scale Indian IT services companies. *International Journal of Green Economics* 4(3): 65–82.

Mohapatra, S. and B. Mohanty. 2001. Defect prevention through defect prediction: a case study at Infosys Technologies Limited. *IECON'01. 2001. Special Session on Software Engineering*, IEEE.

Mohapatra, S. and S. Murarka. 2016. Improving patient care in hospital in India by monitoring influential parameters. *International Journal of Healthcare Management* 9(2): 83–101. http://www.tandfonline.com/doi/full/10.1080/20479700.2015.1101938#abstract.

Mohapatra, S. and R. Rajeev. 2012. Moving to products: entrepreneurial IT firms in India. *International Journal of Project Organization and Management* 4(1): 34–62.

Mohapatra, S., L. Sabharwal, and A. Sud Shabrwal. 2013. Bridge – the missing link. *Asian Case Research Journal* 17(1).

Mohapatra, S., R. Padhi, and S. Mohanty. 2015. Study of motivational factors post-implementing ERP (SAP) solutions in National Aluminium Company Limited extending the technology acceptance model at organisation level (TAMO). *International Journal of Management and Enterprise Development* 14(3): 250–271.

Mori, J. and V. Mahalec. 2016. Planning and scheduling of steel plates production. part ii: Scheduling of continuous casting. *Computers & Chemical Engineering*. In Press.

Mustafa, K. and K. Cheng. 2017. Improving production changeovers and the optimization: A simulation based virtual process approach and its application perspectives. *Procedia Manufacturing* 11: 2042–2050.

Nascimento, M. C. and F. Toledo. 2008. A hybrid heuristic for the multi-plant capacitated lot sizing problem with setup carry-over. *Journal of the Brazilian Computer Society* 14(4): 7–15.

Ogbeide, S. and I. Omorodion. 2016. Development of a model for optimal utilization of raw material in cement production planning process. *Journal of Multidisciplinary Engineering Science and Technology* 3(4): 4437–4442.

Özdamar, L. and G. Barbarosoğlu. 1999. Hybrid heuristics for the multi-stage capacitated lot sizing and loading problem. *Journal of the Operational Research Society* 50(8): 810–825.

Özdamar, L. and Ş. İ. Birbil. 1998. Hybrid heuristics for the capacitated lot sizing and loading problem with setup times and overtime decisions. *European Journal of Operational Research* 110(3): 525–547.

Özdamar, L. and M. A. Bozyel. 2000. The capacitated lot sizing problem with overtime decisions and setup times. *IIE Transactions* 32(11): 1043–1057.

Oztürk, C. and A. Ornek. 2010. Capacitated lot sizing with linked lots for general product structures in job shops. *Computers & Industrial Engineering* 58(1): 151–164.

Park, H., Y. Hong, and S. Y. Chang. 2002. An efficient scheduling algorithm for the hot coil making in the steel mini-mill. *Production Planning & Control* 13(3): 298–306.

Paul, S. M., D. S. Mytelka, C. T. Dunwiddie, C. C. Persinger, B. H. Munos, S. R. Lindborg, and A. L. Schacht. 2010. How to improve r&d productivity: the pharmaceutical industry's grand challenge. *Nature Reviews Drug Discovery* 9(3): 203.

Pochet, Y. and L. A. Wolsey. 1988. Lot-size models with backlogging: Strong reformulations and cutting planes. *Mathematical Programming* 40(1–3): 317–335.

Pochet, Y. and L. A. Wolsey. 2006. *Production planning by mixed integer programming*. Springer Science & Business Media.

Poltroniere, S. C., K. C. Poldi, F. M. B. Toledo, and M. N. Arenales. 2008. A coupling cutting stock-lot sizing problem in the paper industry. *Annals of Operations Research* 157(1): 91–104.

Porkka, P., A. Vepsäläinen, and M. Kuula. 2003. Multiperiod production planning carrying over set-up time. *International Journal of Production Research* 41(6): 1133–1148.

Potts, C. N. and L. van Wassenhove. 1992. Integrating scheduling with batching and lot-sizing: a review of algorithms and complexity. *Journal of the Operational Research Society* 395–406.

Quadt, D. and H. Kuhn. 2003. Production planning in semiconductor assembly. *Proceedings of the Fourth Aegean International Conference on Analysis of Manufacturing Systems*.

Quadt, D. and H. Kuhn. 2009. Capacitated lot-sizing and scheduling with parallel machines, back-orders, and setup carry-over. *Naval Research Logistics (NRL)* 56(4): 366–384.

Ramya, R., C. Rajendran, and H. Ziegler. 2016. Capacitated lot-sizing problem with production carry-over and set-up splitting: mathematical models. *International Journal of Production Research* 54(8): 2332–2344.

Rardin, R. L. and L. A. Wolsey. 1993. Valid inequalities and projecting the multicommodity extended formulation for uncapacitated fixed charge network flow problems. *European Journal of Operational Research* 71(1): 95–109.

Sahling, F., L. Buschkühl, H. Tempelmeier, and S. Helber. 2009. Solving a multi-level capacitated lot sizing problem with multi-period setup carry-over via a fix-and-optimize heuristic. *Computers & Operations Research* 36(9): 2546–2553.

Salomon, M., L. G. Kroon, R. Kuik, and L. N. van Wassenhove. 1991. Some extensions of the discrete lotsizing and scheduling problem. *Management Science* 37(7): 801–812.

Santos, M. O. and B. Almada-Lobo. 2012. Integrated pulp and paper mill planning and scheduling. *Computers & Industrial Engineering* 63(1): 1–12.

Sarkar, S., S. Mohapatra, and J. Sundarakrishnan. 2017. Assessing impact of technology based digital equalizer programme on improving student learning outcomes. *Education and Information Technologies* 22(1): 195–213.

Selen, W. and R. Heuts. 1990. Operational production planning in a chemical manufacturing environment. *European Journal of Operational Research* 45(1): 38–46.

Shim, I.-S., H.-C. Kim, H.-H. Doh, and D.-H. Lee. 2011. A two-stage heuristic for single machine capacitated lot-sizing and scheduling with sequence-dependent setup costs. *Computers & Industrial Engineering* 61(4): 920–929.

Silver, E. A. and H. Meal. 1973. A heuristic for selecting lot size quantities for the case of a deterministic time-varying demand rate and discrete opportunities for replenishment. *Production and Inventory Management* 14(2): 64–74.

Smith-Daniels, V. and L. P. Ritzman. 1988. A model for lot sizing and sequencing in process industries. *The International Journal of Production Research* 26(4): 647–674.

Smith-Daniels, V. L. and D. E. Smith-Daniels. 1986. A mixed integer programming model for lot sizing and sequencing packaging lines in the process industries. *IIE Transactions* 18(3): 278–285.

Sox, C. R. and Y. Gao 1999. The capacitated lot sizing problem with setup carry-over. *IIE Transactions* 31(2): 173–181.

Stadtler, H. 1996. Mixed integer programming model formulations for dynamic multi-item multi-level capacitated lotsizing. *European Journal of Operational Research* 94(3): 561–581.

Stadtler, H. 1997. Reformulations of the shortest route model for dynamic multi-item multi-level capacitated lotsizing. *Operations-Research-Spektrum* 19(2): 87–96.

Stadtler, H. 2000. Improved rolling schedules for the dynamic single-level lot-sizing problem. *Management Science* 46(2): 318–326.

Stadtler, H. 2003. Multilevel lot sizing with setup times and multiple constrained resources: Internally rolling schedules with lot-sizing windows. *Operations Research* 51(3): 487–502.

Stadtler, H. 2005. Supply chain management and advanced planning—-basics, overview and challenges. *European journal of operational research* 163(3): 575–588.

Stadtler, H. and C. Kilger. 2002. *Supply chain management and advanced planning.*

Suerie, C. 2006. *Time continuity in discrete time models: New approaches for production planning in process industries.* Springer Science & Business Media.

Suerie, C. and H. Stadtler. 2003. The capacitated lot-sizing problem with linked lot sizes. *Management Science* 49(8): 1039–1054.

Sung, C. and C. T. Maravelias. 2008. A mixed-integer programming formulation for the general capacitated lot-sizing problem. *Computers & Chemical Engineering* 32(1): 244–259.

Tempelmeier, H. and L. Buschkühl. 2009. A heuristic for the dynamic multi-level capacitated lotsizing problem with linked lotsizes for general product structures. *OR Spectrum* 31(2): 385–404.

Tempelmeier, H. and M. Derstroff. 1996. A lagrangean-based heuristic for dynamic multilevel multiitem constrained lotsizing with setup times. *Management Science* 42(5): 738–757.

Tempelmeier, H. and S. Helber. 1994. A heuristic for dynamic multi-item multi-level capacitated lotsizing for general product structures. *European Journal of Operational Research* 75(2): 296–311.

Thizy, J. and L. N. van Wassenhove. 1985. Lagrangean relaxation for the multi-item capacitated lot-sizing problem: A heuristic implementation. *IIE Transactions* 17(4): 308–313.

Timpe, C. 2002. Solving planning and scheduling problems with combined integer and constraint programming. *OR Spectrum* 24(4): 431–448.

Toledo, C. F. M., M. da Silva Arantes, M. Y. B. Hossomi, and B. Almada-Lobo. 2016. Mathematical programming-based approaches for multi-facility glass container production planning. *Computers & Operations Research* 74: 92–107.

Toledo, C. F. M., M. da Silva Arantes, M. Y. B. Hossomi, P. M. França, and K. Akartunalı. 2015. A relax-and-fix with fix-and-optimize heuristic applied to multi-level lot-sizing problems. *Journal of Heuristics* 21(5): 687–717.

Trigeiro, W. W., L. J. Thomas, and J. O. McClain. 1989. Capacitated lot sizing with setup times. *Management science* 35(3): 353–366.

Ventura, J. A., D. Kim, and F. Garriga. 2002. Single machine earliness–tardiness scheduling with resource-dependent release dates. *European Journal of Operational Research* 142(1): 52–69.

Wagelmans, A., S. van Hoesel, and A. Kolen. 1992. Economic lot sizing: an o (n log n) algorithm that runs in linear time in the wagner-whitin case. *Operations Research* 40(Supplement 1: Optimization): S145–S156.

Wagner, H. M. and T. M. Whitin. 1958. Dynamic version of the economic lot size model. *Management Science* 5(1): 89–96.

Wilson, R. 1934. A scientific routine for stock control. *Harvard Business Review* 13(1): 116–128.

Wolsey, L. 1989. Strong formulations for mixed integer programming: a survey. *Mathematical Programming* 45(1): 173–191.

Wolsey, L. A. 1995. Progress with single-item lot-sizing. *European Journal of Operational Research* 86(3): 395–401.

Wu, T., K. Akartunalı, J. Song, and L. Shi. 2013. Mixed integer programming in production planning with backlogging and setup carryover: modeling and algorithms. *Discrete Event Dynamic Systems* 23(2): 211–239.

Wu, T. and L. Shi. 2011. Mathematical models for capacitated multi-level production planning problems with linked lot sizes. *International Journal of Production Research* 49(20): 6227–6247.

Zabel, E. 1964. Some generalizations of an inventory planning horizon theorem. *Management Science* 10(3): 465–471.

Zangwill, W. I. 1966. A deterministic multi-period production scheduling model with backlogging. *Management Science* 13(1): 105–119.

Zangwill, W. I. 1969. A backlogging model and a multi-echelon model of a dynamic economic lot size production system-a network approach. *Management Science* 15(9): 506–527.

MIX
Papier aus verantwortungsvollen Quellen
Paper from responsible sources
FSC® C105338

If you have any concerns about our products,
you can contact us on
ProductSafety@springernature.com

In case Publisher is established outside the EU,
the EU authorized representative is:
Springer Nature Customer Service Center GmbH
Europaplatz 3, 69115 Heidelberg, Germany

Printed by Libri Plureos GmbH
in Hamburg, Germany